Advanced Diffusion Studies of

Active Enzymes and Nanosystems

Advanced Diffusion Studies of Active Enzymes and Nanosystems

Von der Fakultät Chemie der Universität Stuttgart

zur Erlangung der Würde eines Doktors der

Naturwissenschaften (Dr. rer. nat.) genehmigte Abhandlung

Vorgelegt von

Jan-Philipp Günther

aus Heidelberg

Hauptberichter:	Prof. Dr. Peer Fischer
Mitberichter:	Apl. Prof. Dr. Thomas Sottmann
Mitprüfer:	Prof. Dr. Bernhard Hauer
Tag der mündlichen Prüfung:	15. Dezember 2020

Max-Planck-Institut für Intelligente Systeme, Stuttgart

Institut für Physikalische Chemie, Universität Stuttgart

Bibliografische Information der Deutschen Nationalbibliothek

Die Deutsche Nationalbibliothek verzeichnet diese Publikation in der Deutschen Nationalbibliografie; detaillierte bibliographische Daten sind im Internet über http://dnb.d-nb.de abrufbar.

1. Aufl. - Göttingen: Cuvillier, 2021

Zugl.: (MPI) Stuttgart, Univ., Diss., 2020

Nonnenstieg 8, 37075 Göttingen

Telefon: 0551-54724-0

Telefax: 0551-54724-21

www.cuvillier.de

1. Auflage, 2021

Gedruckt auf umweltfreundlichem, säurefreiem Papier aus nachhaltiger Forstwirtschaft.

ISBN 978-3-7369-7364-0

eISBN 978-3-7369-6364-1

To my family

"Love is just a chemical. We give it meaning by choice."[6]

Eleanor Lamb

Table of Contents

Abbreviations

3HT	His-tag modified bacteriophages (pIII coat protein)
ABEL	anti-Brownian electrophoretic
ACF	autocorrelation function
ALD	fructose-bisphosphate aldolase
AMPPNP	β,γ-imidoadenosine-5'-triphosphate
APD	avalanche photo diode
ATP	adenosine triphosphate
b	diffusion weighting factor
$\boldsymbol{B}$	magnetic field
B_0	static magnetic field strength
B_1	oscillating magnetic field strength
D	(translational) diffusion coefficient
D_0	diffusion coefficient in absence of interactions
D_{app}	apparent diffusion coefficient
D_{eff}	effective diffusion coefficient
D_{rot}	rotational diffusion coefficient
δ	chemical shift
δ_G	gradient pulse length
Δ	diffusion time
ΔE	energy difference
$\Delta \boldsymbol{x}^2$	mean-square-displacement
DHAP	dihydroxyacetone phosphate
DLS	dynamic light scattering
DNA	deoxyribonucleic acid
E-P-C	enzyme-phage-colloid
η	viscosity
$\boldsymbol{F}$	force
FBP	fructose-1,6-bisphosphate
FCCS	fluorescence cross correlation spectroscopy
FCS	fluorescence correlation spectroscopy
FID	free induction decay
FPLC	fast protein liquid chromatography

FRET	Förster resonance energy transfer
G	autocorrelation function (FCS)
G	gradient-pulse strength (NMR)
G3P	glyceraldehyde 3-phosphate
GDH	α-glycerophosphate dehydrogenase
γ	gyromagnetic ratio
h	Planck constant
$\boldsymbol{I}$	angular momentum
I	spin quantum number (NMR theory)
I	fluorescence/echo signal intensity (FCS/NMR)
I_0	signal intensity in absence of quencher/gradient (FCS/NMR)
ISC	intersystem crossing
k	process rate
$\mathrm{k_B}$	Boltzmann constant
k_{cat}	catalytic reaction rate
L	length of a swimmer
λ	boundary layer thickness (phoretic theory)
λ	wavelength (optical spectroscopy)
Λ	combined interaction parameter
$\boldsymbol{M}$	macroscopic magnetization
$\boldsymbol{M}_0$	equilibrium magnetization
m_I	magnetic quantum number
μ	individual magnetic moment
μ	Stokes drag coefficient
N	average number of fluorophores
∇	nabla operator
NADH	nicotinamide adenine dinucleotide
NHS	N-hydroxysuccinimide
NMR	nuclear magnetic resonance
NPP	p-nitrophenylphosphate
ν_0	Larmor frequency
ω_0	radius in radial direction
p	pressure
P	product
PAGE	polyacrylamide gel electrophoresis

PBS	phosphate-buffered saline
PDB	protein data bank
PFG-NMR	pulsed field gradient-NMR
PP	pyrophosphate
ψ	interaction potential
Q	quencher molecule
R	radius
R_h	hydrodynamic radius
Re	Reynolds number
RF	radio frequency
ρ	density
S	substrate (reactant)
S	singlet state (Jablonski diagram)
SDS	sodium dodecyl sulfate
SPT	single particle tracking
ss	single-stranded
STED	stimulated emission depletion
sulfo-SIAB	sulfosuccinimidyl(4-iodoacetyl)aminobenzoate
T	absolute temperature
T	fraction of molecules in the triplet state (FCS)
t	time
T_1	spin-lattice relaxation constant
T_2	spin-spin relaxation constant
τ	lag time
τ_D	translational diffusion time
τ_F	fluorescence lifetime
τ_R	correlation time (NMR relaxation)
τ_T	triplet relaxation time
TCSPC	time-correlated single photon counting
TPI	triosephosphate isomerase
Tris-d_{11}	tris(hydroxymethyl-d_3)amino-d_2-methane
$\boldsymbol{u}$	mass flow velocity
v	swimming speed
z_0	radius in axial direction

Abstract

Enzymes are fascinating nanomachines, which catalyze the reactions essential for life. Studying enzymes is therefore important in a biological and medical context, but the catalytic potential of enzymes also finds use in organic synthesis. This thesis is concerned with the fundamental question whether the catalytic reaction of an enzyme can cause it to show enhanced diffusion. Additionally, this thesis examines if it is possible for enzymes to collectively affect fluid flow, when they are incorporated in functional biohybrid nanostructures.

The diffusive behavior of enzymes is important for their distribution within cells and impacts their ability to reach their substrates. Several reports within the last decade have claimed that enzymes show diffusion enhancements of up to 80 %, when they are catalytically active. Examining these claims is important to achieve an understanding of their biological function. These reports were based on fluorescence correlation spectroscopy (FCS), which measures the fluorescence fluctuations of individual fluorophores (here, labeled enzymes) passing through a small focal volume. FCS is a powerful tool to study diffusion, but several photo- and biophysical processes can interfere with FCS measurements. The work presented in this thesis made theoretical predictions how these effects can lead to misinterpretations specific to FCS experiments of active enzymes. Additionally, these simulations were supported by multi-detector FCS experiments, which showed that the 80 % diffusion enhancement reported by others is actually a misinterpretation of a complex fluorescence quenching artefact. The FCS experiments reported within this thesis find no evidence for active enzyme diffusion enhancement.

To completely rule out the possibility of fluorescence artefacts, another diffusion measurement technique, which does not require labeling, was adapted for enzymes: Diffusion nuclear magnetic resonance (NMR) spectroscopy. Hence, the work presented in this thesis reports the first diffusion NMR experiment of an active enzyme. Earlier reports by others claim that enzymes, which catalyze endothermic reactions, self-propel and cause enhanced diffusion of molecules in their surroundings. The diffusion NMR studies in this thesis reveal that neither the enzymes themselves nor tracer molecules in the solution show enhanced diffusion.

These findings of enzyme diffusion NMR measurements and the unraveling of artefacts in FCS enzyme measurements seriously question the hypothesis that enzymes are active matter and experience enhanced diffusion. The publication of these results was the first to experimentally question this hypothesis. Since then, several reports by others have appeared that support the predictions and experimental observations presented herein.

In addition to the academic interest, there is also a practical interest in enzymes due to their fascinating properties as catalysts. Biocatalysis, which uses enzymes or whole organisms for organic synthesis, has several advantages over conventional catalysis. Enzymes are, for instance, highly selective and efficient, can be operated in mild conditions and are safe to dispose of. The natural enantioselectivity of enzymes in some reactions is of special interest for the synthesis of pharmaceuticals, which are often homochiral. However, reuse of enzymes by recovery from reaction mixtures is problematic due to their small size and fragility. Immobilization of enzymes onto microparticles simplifies the recovery step, but often lowers the enzymes' catalytic activity. In this thesis, a novel nanoconstruct is presented, which allows easy recovery of enzymes with a magnet, but still ensures high enzymatic activity. In this construct, filamentous viruses are utilized as intermediate immobilization templates between the microparticle and the enzyme. This novel nanoconstruct has been termed enzyme-phage-colloid (E-P-C). Within this thesis two applications for E-P-Cs are presented. The first application is the repeated use of E-P-Cs as biocatalysts with easy magnetic recovery between each reaction cycle. Activity assays showed that the enzyme had even higher catalytic turnover, when it was immobilized on the E-P-C. In the second application, E-P-Cs were used to construct an enzymatic micropump, which is a microdevice that creates convective flows due to density differences with a locally catalyzed reaction. E-P-Cs can conveniently be immobilized with a magnet onto the wall of a microcontainer to form an enzymatic micropump. These E-P-C micropumps are shown to generate the fastest flow speeds of an enzymatic micropump to date. Additionally, it is shown that urease E-P-Cs can pump blood at physiological urea concentrations, which might enable medical lab-on-a-chip applications.

The last part of this thesis considers the diffusion of active molecular catalysts. These synthetic catalysts have recently been reported to show enhanced diffusion in diffusion NMR experiments similar to enzymes. This is particularly surprising as the molecular catalysts exhibit catalytic turnover rates that are orders of magnitude lower than the ones of enzymes. In addition to the proposed self-propulsion, it was claimed that the molecular catalysts transfer kinetic energy to the surrounding solvent and substrate molecules, thereby causing their diffusion enhancement. However, re-examination of these diffusion NMR studies showed that there is no diffusion enhancement of molecular catalysts and the misinterpretation of the diffusion NMR experiments is caused by a complex artefact due to intensity changes over the course of the diffusion experiment, which are caused by relaxation phenomena. These results have recently been submitted for publication.

Kurzzusammenfassung

Enzyme sind faszinierende Nanomaschinen und katalysieren alle Reaktionen, die essentiell für das Leben sind. Die Forschung an Enzymen ist daher notwendig für die Biologie und Medizin, aber die katalytischen Fähigkeiten von Enzymen finden auch Anwendung in der organischen Synthese. Diese Arbeit befasst sich mit der Frage, ob die enzymatische Katalyse bei Enzymen selbst erhöhte Diffusion hervorrufen kann. Außerdem untersucht sie, ob es möglich ist für Enzyme kollektiv Strömungen in Flüssigkeiten hervorzurufen, wenn sie in funktionale Biohybrid-Nanostrukturen eingebunden sind.

Das Diffusionsverhalten von Enzymen ist wichtig für ihre Verteilung in Zellen und beeinflusst die Fähigkeit von Enzymen Substrate zu erreichen. Mehrere Berichte innerhalb des letzten Jahrzehnts behaupten, dass Enzyme ihre Diffusion um bis zu 80 % erhöhen, wenn sie katalytisch aktiv sind. Dies hätte große Auswirkungen auf unser Verständnis der biologischen Funktion von Enzymen. Die erwähnten Berichte basieren auf Messungen der Fluoreszenzkorrelationsspektroskopie (FCS), bei der die Fluoreszenzfluktuationen einzelner Fluorophore (hier, fluoreszenzmarkierte Enzyme) gemessen werden, während diese sich durch ein kleines fokales Volumen bewegen. FCS ist eine wertvolle Methode für Diffusionsmessungen, wird aber von foto- oder biophysikalischen Prozessen beeinflusst. Diese Arbeit beinhaltet theoretische Voraussagen darüber wie diese Prozesse zu Fehlinterpretationen speziell im Fall von FCS Messungen aktiver Enzyme führen können. Zusätzlich wurden diese Simulationen von Multidetektor-FCS Messungen bestätigt, was zu der Erkenntnis führte, dass die berichteten 80 % Diffusionserhöhung tatsächlich auf einer Fehlinterpretation eines komplexen Fluoreszenzlöschungsartefakts basieren. Die FCS Messungen in dieser Arbeit zeigen daher keine Diffusionserhöhung aktiver Enzyme.

Um Fluoreszenzartefakte komplett ausschließen zu können, wurde eine weitere Methode zur Diffusionsmessung, welche keine Probenmarkierung benötigt, auf Enzymmessungen erweitert. Hierbei handelt es sich um die Methode der Diffusionskernspinresonanz-spektroskopie (Diffusions-NMR-Spektroskopie). Diese Arbeit beinhaltet daher die ersten Diffusions-NMR-Messungen aktiver Enzyme. Frühere Arbeiten anderer Forschungsgruppen berichten von erhöhter Diffusion von Enzymen, die endotherme Reaktionen katalysieren, was außerdem angeblich zu erhöhter Diffusion von Molekülen in der Umgebung des Enzyms führen soll. Diffusions-NMR-Messungen in dieser Arbeit zeigen jedoch weder für Enzyme selbst noch für Moleküle in der Reaktionslösung Diffusionserhöhung.

Diese NMR-Ergebnisse zur Enzymdiffusion und die Entdeckung von Artefakten in FCS-Messungen von Enzymen widersprechen der Hypothese, dass Enzyme aktive Materie sind und aktiv ihre Diffusion erhöhen. Die Veröffentlichungen dieser Ergebnisse stellen als erste experimentelle Studien diese Hypothese in Frage. Weitere Publikationen anderer Forschungsgruppen untermauern die hier präsentierten Ergebnisse.

Zusätzlich zu dem akademischen Interesse gibt es auch ein wirtschaftliches Interesse an den beeindruckenden katalytischen Fähigkeiten von Enzymen. Die Biokatalyse, welche Enzyme oder vollständige Organismen zur organischen Synthese verwendet, hat mehrere Vorteile gegenüber der konventionellen Katalyse. Enzyme sind zum Beispiel hoch selektiv, effizient, können unter milden Bedingungen arbeiten und sind sicher in der Entsorgung. Die natürliche Entioselektivität von Enzymen in einigen Reaktionen ist von besonderem Interesse für die Synthese pharmazeutischer Wirkstoffe, da diese oft homochiral sind. Jedoch ist die Wiedergewinnung von Enzymen aus Reaktionsmischungen problematisch aufgrund ihrer kleinen Größe und Sensibilität. Immobilisierung von Enzymen hingegen kann die Wiedergewinnung stark vereinfachen, erniedrigt aber oft die Enzymaktivität. Diese Arbeit präsentiert ein neuartiges Nanokonstrukt, welches die einfache Wiedergewinnung von Enzymen mit einem Magneten erlaubt und dabei die katalytische Aktivität des Enzymes erhält. Das Konstrukt besteht aus filamentösen Viren, welche als intermediäre Immobilisierungstemplate Mikropartikel und Enzym verbinden. Dieses neuartige Nanokonstrukt wird als Enzym-Phagen-Kolloid (E-P-C) bezeichnet. Innerhalb dieser Arbeit werden zwei Anwendungen von E-P-Cs demonstriert. Die erste Anwendung ist der wiederholte Einsatz von E-P-Cs als Biokatalysatoren mittels magnetischer Rückgewinnung zwischen den Reaktionszyklen. Aktivitätsmessungen zeigten, dass die katalytische Aktivität der Enzyme sogar erhöht ist, wenn sie auf dem Nanokonstrukt immobilisiert sind. Die zweite Anwendung beinhaltete die Verwendung von E-P-Cs in enzymatischen Mikropumpen, welche

Konvektionsströme durch Dichtegradienten mittels lokaler Katalyse erzeugen. Um eine enzymatische Mikropumpe herzustellen, können E-P-Cs mit einem Magneten einfach an der Wand eines Mikrocontainers lokalisiert werden. Die enzymatischen Mikropumpen, welche auf diese Art hergestellt wurden, erzeugen die schnellsten Flussgeschwindigkeiten, die bis heute in enzymatischen Mikropumpen gemessen wurden. Zusätzlich wurde gezeigt, dass E-P-Cs aus Urease Blut mittels seines natürlichen Harnstoffgehaltes durch enzymatische Mikropumpen bewegen können. Dies könnte in Form von mikrofluidischen Lab-on-a-Chip Systemen Anwendung in der Medizin finden.

Der letzte Teil dieser Arbeit befasst sich mit der Diffusion von aktiven molekularen Katalysatoren. Vor kurzem veröffentlichte Studien basierend auf Diffusions-NMR-Messungen berichten von erhöhter Diffusion dieser synthetischen Katalysatoren in einer ähnlichen Weise wie es für Enzyme vermutet wird. Diese Hypothese ist besonders verwunderlich, da molekulare Katalysatoren um Größenordnungen niedrigere Aktivitäten aufweisen als Enzyme. Zusätzlich zu der vermeintlichen Eigendiffusionserhöhung wird auch von einem Übertrag der kinetischen Energie der molekularen Katalysatoren auf die sie umgebenden Substrat- und Lösungsmoleküle berichtet, was wiederum zu deren Diffusionserhöhung führen soll. Der Teil dieser Arbeit, welcher sich diesem Thema widmet, wurde vor Kurzem zur Veröffentlichung eingereicht. Es war möglich zu zeigen, dass weder die molekularen Katalysatoren noch die Moleküle in ihrer Umgebung Diffusionserhöhung erfahren und dass die Fehlinterpretationen der Diffusions-NMR-Messungen auf komplexen Artefakten basiert, welche durch Signalintensitätsänderungen basierend auf Relaxationsphänomenen während des Diffusionsexperiments hervorgerufen werden.

1 Introduction

Enzymes are nature's smallest machines and essential for life itself. From the production of ATP by ATP synthase in mitochondria to the motion of muscles by myosin all life is controlled by enzymes. These examples are motor proteins, which convert chemical energy into rotational or linear motion during their catalytic cycle. Motor proteins are embedded in membranes or moving along protein fibers to transfer their motion, but the majority of enzymes is freely diffusing in cells or other biological fluids. Several reports within the last decade have claimed that these freely diffusing enzymes may also be able to propel themselves, i.e., swim, when they are catalytically active.[7–10] If this claim would be true, it would change our fundamental understanding of the function and motion of enzymes and therefore would have far-reaching implications for the biology of the cell, pharmaceuticals and medicine. One part of this thesis is therefore dedicated to testing the active enzyme diffusion hypothesis with advanced diffusion measurement techniques.

The initial reports in support of this hypothesis were based on measurements of fluorescence correlation spectroscopy (FCS). This technique, which is discussed in detail in Chapter 4.1, measures the diffusion of fluorophores (here, fluorescently labeled proteins) through the confocal volume in a microscope. The time needed to pass through this volume is therefore a measure of the samples diffusion coefficient. Artefacts, which change the photophysical behavior of the fluorophore or the diffusing sample, can, however, lead to misinterpretations of the otherwise very precise FCS method, as is shown in Chapter 7 (Ref. 1). Therefore, multi-detector FCS measurements together with careful enzyme purification studies were performed within this thesis to thoroughly test the active enzyme diffusion hypothesis. Initially, two enzymes (alkaline phosphatase and F_1-ATPase) were tested and the results are presented in Chapter 7 (Ref. 1).

To avoid any possible photophysical artefacts, a label-free technique was also chosen to test the active enzyme diffusion hypothesis. Therefore, diffusion nuclear magnetic resonance (NMR) measurements were introduced within this thesis as an alternative to FCS measurements to carefully measure the diffusion of enzymes and solvent molecules. This is demonstrated for the enzyme aldolase (see Chapter 8, Ref. 2), which is of particular interest, since it has been reported[9] to show enhanced diffusion even though no heat is released during its catalyzed reaction. All experiments presented in Chapter 7 and 8 (Ref. 1 and 2) found no evidence of enhanced enzyme diffusion. Rather, it was possible to identify measurement errors in earlier publications by other groups,[11,8,9,12] which had reported enhanced enzyme diffusion. The implications of the results presented in Chapter 7 and 8 (Ref. 1 and 2) for the field of enhanced enzyme diffusion are summarized in Chapter 11.

In addition to the academic interest in enzymes for a deeper understanding of biochemical processes, there is also interest in enzymes as biocatalysts for applications in synthesis. This is due to several advantages that enzymes have over conventional chemical catalysts (see Chapter 6.1). Enzymes are highly selective and efficient catalysts, which can be operated in mild conditions and are safe to dispose of, just to name a few advantages. The natural enantioselectivity of enzymes in some reactions is of special interest for pharmaceuticals which are often homochiral. In recent decades, biocatalysis has reached increasing importance due to novel efforts in directed enzyme evolution, which increases the spectrum of reactions enzymes can catalyze. However, separation of enzymes from the reaction mixture is problematic due to their small size and fragility, but their recovery is of economic interest. Immobilization of enzymes onto microparticles simplifies their recovery, but often lowers their catalytic activity, since the enzyme can be locked in certain conformational positions due to tight binding to a solid template or the supply of substrate his hindered, if the enzyme's active site is facing the surface of the template. In this thesis, a novel nanoconstruct is presented, which allows easy recovery of the enzyme with a magnet, but still ensures high enzyme activity. In this construct, filamentous viruses, which are harmless to humans, have been utilized as templates to immobilize enzymes. The viruses were then anchored onto a magnetic carrier particle. There is no reduction of the enzyme's activity, since virus and enzyme both consist of proteins and since the virus is a much more flexible template than a solid surface and allows sufficient substrate diffusion to the enzyme due to its filamentous structure. This novel nanoconstruct has been termed enzyme-phage-colloid (E-P-C) and is introduced together with two of its applications in Chapter 9 (Ref. 3).

Recently, similar claims to the active enzyme diffusion hypothesis have been made about (synthetic) molecular catalysts.[13] It has been claimed that the transfer of energy from catalysts to substrate and solvent molecules in their surrounding is a common phenomenon in chemical reactions. This would have substantial impact on our understanding of the chemical reaction. So far it was believed that energy released by chemical reactions (including catalyzed reactions) is dissipated on small length scales and fast timescales via vibrational energy relaxation processes. The authors of Ref. 13 suggest that a substantial amount of energy from the reaction is transferred into enhanced diffusion without significant temperature change of the solution, which seems unphysical since the molecules move within the highly overdamped low Reynolds number regime (Chapter 2.1). Additionally, the claims in Ref. 13 are especially extraordinary, since molecular catalysts have orders of magnitude lower catalytic turnover rates than enzymes, which leads to a much lower power output of the catalyst. Since the enhanced diffusion of enzymes has been deemed thermodynamically impossible by some,[14] this would make enhanced diffusion for molecular catalysts even more unlikely. Nevertheless, it is necessary to find an explanation for the observations in Ref. 13. Therefore, the experiments in Ref. 13 have been reproduced as part of this thesis, but with thorough controls, which could reveal an entirely different explanation, which is in fact not at all connected to active diffusion, but to intensity changes due to relaxation phenomena during the NMR experiment (Chapter 10, Ref. 4).

This thesis is structured as follows. First, the fundamental aspects of motion at the nanoscale are described in Chapter 2. This includes the requirements for swimming at low Reynolds numbers and a description of Brownian motion. Active motion at the nanoscale is then discussed in Chapter 3, where self-phoresis is introduced. Additionally, Chapter 3 includes an overview of the experimental and theoretical reports on the active enzyme diffusion hypothesis and active molecular catalyst diffusion. The advanced diffusion measurement techniques, which are used in this thesis, are presented in detail in Chapter 4, which mainly includes FCS and diffusion NMR. Detailed knowledge of enzyme properties is needed for enzyme diffusion measurements and to utilize enzymes in biocatalysis. Therefore, the properties of the enzymes studied in this thesis are presented in Chapter 5. How to utilize and immobilize enzymes is discussed in Chapter 6. These introductory chapters are followed by the published experimental results (Refs. 1–4) which are briefly summarized and reprinted in Chapters 7-10.

1.1 Publications

This cumulative thesis is based on the following publications (published Refs. 1–3, and submitted Ref. 4):

- J.-P. Günther, M. Börsch, and P. Fischer
  Diffusion Measurements of Swimming Enzymes with Fluorescence Correlation Spectroscopy
  Acc. Chem. Res. **51**, 1911–1920 (2018)

- J.-P. Günther, G. Majer, and P. Fischer
  Absolute Diffusion Measurements of Active Enzyme Solutions by NMR
  J. Chem. Phys. **150**, 124201 (2019)

- M. Alarcón-Correa,* J.-P. Günther,* J. Troll, V. M. Kadiri, J. Bill, P. Fischer, and D. Rothenstein*
  Self-Assembled Phage-Based Colloids for High Localized Enzymatic Activity
  ACS Nano **13**, 5810–581 (2019)

- J.-P. Günther,* L. L. Fillbrook,* T. S. C. MacDonald, G. Majer, W. S. Price, P. Fischer, and J. E. Beves
  Comment on "Boosted molecular mobility during common chemical reactions"
  ChemRxiv Preprint, DOI: 10.26434/chemrxiv.13023164 (2020)

Other work published by the author during his PhD program:

- V. M. Kadiri, M. Alarcón-Correa, J. Ruppert, J.-P. Günther, J. Bill, D. Rothenstein, and P. Fischer
  Genetically Modified M13 Bacteriophage Nanonets for Enzyme Catalysis and Recovery
  Catalysts **9**, 723 (2019)

- H.-H. Jeong,* M. C. Adams,* J.-P. Günther, M. Alarcón-Correa, I. Kim, E. Choi, C. Miksch, A. F. Mark, A. G. Mark, and P. Fischer
  Arrays of Plasmonic Nanoparticle Dimers with Defined Nanogap Spacers
  ACS Nano **13**, 11453–11459 (2019)

- J. Sachs,* J.-P. Günther,* A. G. Mark, and P. Fischer
  Chiroptical Spectroscopy of a Freely Diffusing Single Nanoparticle
  Nat. Commun. **11**, 4513 (2020)

*indicates equal contribution

In the meantime, an updated version of the comment (Ref. 4) has been peer-reviewed and published (Ref. 5):

- J.-P. Günther,* L. L. Fillbrook,* T. S. C. MacDonald, G. Majer, W. S. Price, P. Fischer, and J. E. Beves
  Comment on "Boosted molecular mobility during common chemical reactions"
  Science **371**, eabe8322 (2021)

*indicates equal contribution

2 Motion at the Nanoscale

"What does it mean to swim? Well, it means simply that you are in some liquid and are allowed to deform your body in some manner. That's all you can do. Move it around and move it back."[15]

Edward Mills Purcell

2.1 Motion at Low Reynolds Numbers

Motion in fluids varies drastically over various length scales, due to different forces dominating the dynamics. The motion of macroscopic objects is dominated by inertia, whereas drag dominates on extremely small scales. The latter is known as the low Reynolds number regime. An example for motion at low Reynolds numbers is the self-propulsion of bacteria, which rotate their thin flagella in a helical motion. On a macroscopic scale this may seem inefficient, but in the microscopic world this propulsion mechanism yields speeds of many body lengths per second. The following section aims to explain the fundamental differences of motion at low Reynolds numbers and the underlying physical principles.

Motion of fluids in general (large and small scale) can be described by the Navier-Stokes Equation, which balances the various forces acting on a volume element of a fluid:[16]

$$\rho \frac{D\boldsymbol{u}}{Dt} = -\nabla p + \eta \nabla^2 \boldsymbol{u} + \rho \boldsymbol{F} \tag{2.1}$$

Here, ρ is the density, $\boldsymbol{u}$ the mass flow velocity, $\frac{D}{Dt} = \frac{\partial}{\partial t} + \boldsymbol{u}\nabla$ the material derivative, p the pressure, η the viscosity, and ∇ the nabla operator. The terms of the equation express pressure forces $-\nabla p$, viscous forces $\eta \nabla^2 \boldsymbol{u}$, and external forces $\rho \boldsymbol{F}$ (such as gravity) acting per volume element. This form of the Navier-Stokes Equation can be used for incompressible fluids on any length scale, but we will now try to simplify it for motion on small scales.

The dimensionless Reynolds Number Re is used to estimate the flow conditions of a system. Re is a measure of the relation of inertia to viscous forces and therefore helps to identify different regimes of fluid flow. Re is defined as follows:

$$\mathrm{Re} = \frac{\boldsymbol{F}_{\text{inertia}}}{\boldsymbol{F}_{\text{viscous}}}, \tag{2.2}$$

where $\boldsymbol{F}_{\text{inertia}}$ represents inertia and $\boldsymbol{F}_{\text{viscous}}$ represents viscosity from the Navier-Stokes equation. These forces can be expressed in dependence of the length scale of an object L:

$$\boldsymbol{F}_{\text{inertia}} = \rho \frac{D\boldsymbol{u}}{Dt} \propto \frac{\rho u^2}{L} \tag{2.3}$$

$$\boldsymbol{F}_{\text{viscous}} = \eta \nabla^2 \boldsymbol{u} \propto \frac{\eta u}{L^2} \tag{2.4}$$

From which the dimensionless Re is obtained:

$$\mathrm{Re} \equiv \frac{\rho u L}{\eta} \tag{2.5}$$

Whereas $\mathrm{Re} \gg 1$ corresponds to turbulent flows with dominant inertia, $\mathrm{Re} \ll 1$ represents laminar flows with dominant viscous forces. Macroscopic swimmers such as fish or humans ($\mathrm{Re} \approx 10^2$ to 10^5) use reciprocal beating patters to propel themselves. This is possible, since they can glide during their recovery stroke due to inertia. As will be discussed next, these swimming mechanisms are not possible at $\mathrm{Re} \ll 1$.

At low Re, the Navier-Stokes equation (2.1) can be simplified to yield the Stokes equation, since inertia ($\rho \frac{D\boldsymbol{u}}{Dt} = 0$) and external forces ($\rho \boldsymbol{F} = 0$) can be neglected:[15]

$$-\nabla p + \eta \nabla^2 \boldsymbol{u} = 0 \tag{2.6}$$

The Stokes equation does not contain any explicit time dependency, which implies that all events at $\mathrm{Re} \ll 1$ are time-reversible. As Edward Purcell has pointed out in his article "Life at low Reynolds number",[15] this has several implications on the swimming mechanism employed by microscopic life forms. The reciprocal beating patterns, which are typical for the macroscopic world, cannot propel microorganisms ($\mathrm{Re} \approx 10^{-5}$ for bacteria), since these patterns are symmetric in time and rely on inertia. Therefore, microorganisms have to rely on more complex beating patterns. This has been coined the "scallop theorem",[15] since the reciprocal opening and closing of a scallop does suffice for propulsion at high Re, but not at low Re. Possible strategies to solve the problem of propulsion at low Re are the helical rotation used by *e.g.* bacteria via flegella[17] and travelling waves used by *e.g.* ciliates[18] and sperm.[19] This motion can be compared to the drilling motion of a helical screw and is very effective at low Re. For objects even smaller than individual cells the conditions are even more extreme (*e.g.* motor proteins with $\mathrm{Re} \approx 10^{-8}$) and thermal fluctuations also factor in, as we will discuss in the following section.

2.2 Brownian Motion

Everything is in motion on a molecular scale, since thermal energy is present in all samples above 0 K. This phenomenon has been termed Brownian motion, which was named after Robert Brown, who observed the random motion of small particles from the inside of pollen. Brown did not observe the motion of the pollen themselves as is often erroneously stated, since they are too big to show substantial Brownian motion.[20] The random thermal noise of the solvent is transferred to a small particle by many collisions. Over a longer timescale those collisions result in a random walk of the particle or molecule. This transition from ballistic motion to the random walk has recently been observed with ultrafast video tracking for dispersed particles.[21]

Brownian motion is a form of diffusion. The term "diffusion" is used in the scientific literature to describe multiple phenomena.[22] One is the so-called self-diffusion, which is observed in homogenous samples and mixtures. Here, the root-mean-squared displacement $\langle \Delta \boldsymbol{x}^2 \rangle$ of the substance of interest is measured over time. This can be expressed as a (self-)diffusion coefficient D

$$D = \lim_{t \to \infty} \frac{1}{6t} \langle \Delta x^2 \rangle, \tag{2.7}$$

which is defined for time periods t, which are sufficiently long to average over the random motion. The self-diffusion coefficient usually differs from the mutual diffusion coefficient. The latter describes the diffusion in inhomogeneous samples, where differences in chemical potential lead to mass fluxes. Though mutual diffusion has been studied extensively in the context of enzymes over the last years,[12,23–25] this thesis is focused on self-diffusion. The reason for this is to minimize the effects at play. Mutual diffusion can occur independently of enzyme activity, since only the presence of a substrate gradient is sufficient for mutual diffusion. A focus on self-diffusion (from now on just 'diffusion') therefore helps to better understand, which effects are caused by the enzymes' activity.

Diffusion coefficients of samples in a liquid consisting of solvent molecules, which are small in relation to the solute, can be estimated with the Stokes-Einstein-Sutherland equation[26,27]

$$D = \frac{\mathrm{k_B} T}{\mu}, \tag{2.8}$$

where $\mathrm{k_B}$ is the Boltzmann constant and T the absolute temperature. Since the system is on average force free, hydrodynamic forces and thermal forces are balanced. The hydrodynamic forces contribute to the Stokes-Einstein-Sutherland equation via the drag coefficient μ from Stokes' law, which is dependent on the geometry of the sample and viscosity of the medium. For a sphere with hydrodynamic radius R_h the drag coefficient is $\mu_\mathrm{trans} = 6\pi\eta R_\mathrm{h}$ under the assumption of infinite dilution (self-diffusion coefficient) and stick boundary conditions between the sample and the medium.

In addition to translational diffusion, molecules and particles also undergo rotational diffusion. Rotational diffusion occurs on much shorter timescales than translational diffusion for small objects. The rotational drag coefficient of a sphere is $\mu_\mathrm{rot} = 8\pi\eta {R_\mathrm{h}}^3$.

3 Active Brownian motion

"All swimming ultimately can be traced to the dynamics of enzymes. Both animal muscles and the cytoskeletons of individual cells move through the cooperation of many motor proteins, which are enzymes that act together in large scale structures. Can individual enzymes, however, swim? How would they do so?"[28]

Xiaoyu Bai and Peter Guy Wolynes

The previous chapter described the passive motion of small samples due to thermal fluctuations, but active matter can also exhibit more complex behavior often termed active Brownian motion or phoretic self-propulsion. This usually involves catalytic activity of the sample, where parts of the Gibbs free energy of the reaction is used for (directed) propulsion. Other types of energy input are also possible. Active Brownian motion has been observed over various length scales with very different propulsion mechanisms, which will be discussed in the following.

3.1 Self-Phoresis of Micro- and Nanoparticles

Symmetry breaking, which is required for swimming at low Re as discussed in Chapter 2, can also be achieved by asymmetric coating of microspheres. These structures are called Janus particles in reminiscence of the two-faced Roman god. If one side of the asymmetric coating is catalytically active, the particle is able to undergo a self-phoretic propulsion mechanism. Phoresis describes the migration of a particle (or micro-organism) due to a gradient in its surrounding. In the case of Janus particles this gradient can be created by the particle itself, hence the mechanism is termed self-phoresis. One distinguishes between diffusiophoresis for neutral solute gradients, electrophoresis for gradients of charged species and thermophoresis for temperature gradients. Since most reactions occur in water, diffusiophoresis and electrophoresis generally have to be considered together.[29] Therefore, pure self-diffusiophoresis without charged species might be a purely theoretical concept, but offers many important insights into the propulsion mechanism and scaling of the propulsion speed of Janus swimmers.

Self-diffusiophoresis is possible due to different interactions between the solutes (substrate S and product P) and both sides of the Janus particle. The interactions are usually assumed to be short ranged and only act within a boundary layer with the distance λ from the surface of the particle. When a Janus particle is suspended in a solution containing the substrate, product is formed on the catalytically active half C of the Janus particle (Fig. 3.1). This causes a concentration gradient of product molecules along the particle's surface. If the two halves of the Janus particle have different interaction potentials ψ with the product molecule (or the substrate molecule), a force resulting from the solute gradient is generated. Since the overall system is force-free, a flow is generated, which is counteracted by the propulsion of the particle in the opposite direction to the induced flow.[29,30]

The steady-state swimming speed v of a Janus swimmer can be determined analytically for a spherical Janus swimmer with one catalytically active half C and a noncatalytic half N. v is indicating the speed in the direction from N to C (Fig. 3.1). In the reaction-controlled limit, which is mainly important for small swimmers, the factors influencing v are the following:[30,31]

$$v \propto \frac{\mathrm{k_B}T}{\eta} \frac{[\mathrm{S}]_0}{D} \left({\Lambda_\mathrm{C}}^2 + {\Lambda_\mathrm{N}}^2\right) \frac{k_\mathrm{cat}}{R^2} \tag{3.1}$$

$[\mathrm{S}]_0$ is the substrate concentration before the start of the reaction, D is the diffusion coefficient of the solutes (S and P), R is the radius of the Janus swimmer, ${\Lambda_\mathrm{C}}^2$ and ${\Lambda_\mathrm{N}}^2$ are the interaction parameters between the two interfaces and the solutes, and k_cat is the catalytic reaction rate of the Janus swimmer (here, $k_\mathrm{cat} \propto R^2$). ${\Lambda_\mathrm{C}}^2$ and ${\Lambda_\mathrm{N}}^2$ are dependent on the interaction potentials $\psi_\mathrm{PI}(r)$ and $\psi_\mathrm{SI}(r)$ between the solutes and the interfaces ($\mathrm{I} =$ C or N):[30]

$${\Lambda_\mathrm{I}}^2 = \int_0^\infty dr\, r \left(e^{-\frac{\psi_\mathrm{PI}(r)}{\mathrm{k_B}T}} - e^{-\frac{\psi_\mathrm{SI}(r)}{\mathrm{k_B}T}}\right) \qquad \mathrm{I} \in [\mathrm{C}, \mathrm{N}] \tag{3.2}$$

If $\psi_{PN} = \psi_{SN}$, ${\Lambda_N}^2 = 0$ and the sign of v is decided by ${\Lambda_C}^2$. If the product is more attracted to C than the substrate, then ${\Lambda_C}^2 > 0$ and the Janus swimmer shown in Fig. 3.1 moves to the right. These relationships will later be used to understand the scaling of self-diffusiophoresis with the size of the swimmer.

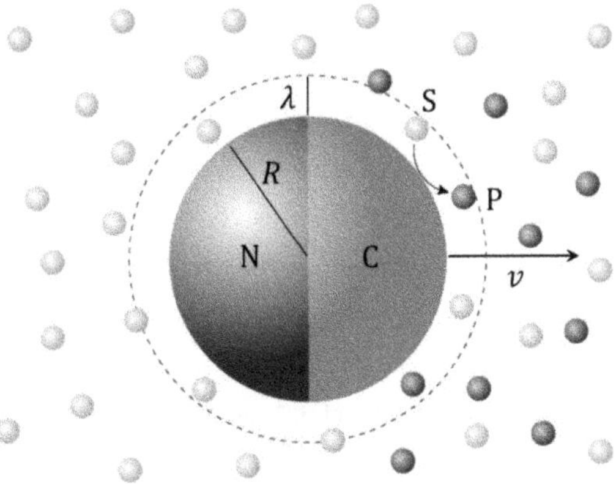

Figure 3.1. Self-diffusiophoresis of a spherical Janus swimmer. The swimmer has a catalytic C (red) and noncatalytic N (blue) half. On C the substrate S (light grey) is converted to the product P (dark grey) within the boundary layer λ.

It can be concluded from Eq. (3.1) that v is independent of the radius R under the assumption that k_{cat} is proportional to the surface area of the swimmer ($k_{cat} \propto R^2$). But with decreasing size of the swimmer, reorientation due to rotational diffusion plays an increasing role and this is expected to reduce the speed. This effect can be estimated by calculating the effective translational diffusion coefficient D_{eff}:[30]

$$D_{eff} = D_0 + \frac{1}{6}\frac{v^2}{D_{rot}} \tag{3.3}$$

D_0 is the translational diffusion coefficient of the swimmer in absence of catalysis and D_{rot} is the rotational diffusion coefficient. For smaller swimmers, the diffusive term is increasing, since $D_0 \propto R^{-1}$, but the propulsion term decreases rapidly, since $v^2/D_{rot} \propto R^3$. Therefore it can be concluded that the motion of small self-diffusiophoretic swimmers is mainly controlled by diffusive motion. A notable exception are swimmers with a catalytic activity at the diffusion-controlled limit, since this changes Eq. (3.1) to yield an overall proportionality of $v \propto R^{-1}$ and therefore $v^2/D_{\mathrm{rot}} \propto R$.[30] In this case the transition between directed and purely diffusive motion is observed for smaller swimmers.

Early examples of self-phoretic Janus swimmers were rod-shaped particles, which were 2 µm in size and reached speeds of 8 µm/s,[32] but more studies followed with increasingly smaller Janus particles.[33–35] Most of these experimental systems relied on the decomposition of hydrogen peroxide on a noble metal surface (*e.g.* Pt). The smallest swimmers realized this way are smaller than 50 nm in diameter.[35,36] The propulsion of Janus swimmers can be treated as a linear propulsion on top of the Janus particles' Brownian diffusion.[37] Therefore, the terms 'enhanced diffusion' or 'active diffusion' are also used in the context of self-phoretic nanoswimmers. In the presence of H_2O_2 the diffusion of 15 and 30 nm Pt/Au Janus particles is increased by 30 and 50 % respectively.[35,36] These results are surprisingly large and were not expected by theoretical estimations of the scaling of self-electrophoresis. Detailed theoretical studies of self-electrophoresis suggest a strong decrease of the propulsion speed for swimmers smaller than 1 µm,[38] whereas theoretical estimations of self-diffusiophoresis still expect significant propulsion even for Ångström-sized swimmers.[31]

Janus swimmers can also be propelled with enzymes instead of inorganic catalysts. This has led to many realizations of enzyme-powered microswimmers in recent years.[39] The smallest enzyme-powered Janus swimmer is only 90 nm in diameter and has been powered with catalase, which has one of the fastest turn-over rates of all enzymes.[34] The structural basis of the enzyme-powered Janus swimmer can be *e.g.* a solid particle,[40,41] a porous particle,[42,43,34] or soft materials such as cells.[44] The advantage of enzyme propulsion over inorganic catalysts is the biocompatibility of fuel, product and catalysts. Many enzymes use biocompatible substrates and are, as proteins themselves, biodegradable. Therefore many enzyme-powered microswimmers were developed with medical applications in mind.[44,45] Those advantages can also be found for enzyme-phage-colloids, which will be discussed in Chapter 9.

3.2 Active Diffusion Hypothesis for Enzymes

Individual enzymes share many similarities with the examples of self-phoretic materials discussed above. Enzymes are catalytically active and enzymes are asymmetric with the active site usually positioned off-center. However, there are also differences, as enzymes can undergo time-asymmetric shape change cycles due to conformational changes in their catalytic cycle.[46] The latter is a requirement for swimming via body-shape changes at low Re. It is therefore interesting to examine the possibility that enzymes themselves could act as self-phoretic matter. The main difference to other self-phoretic particles is that enzymes are roughly one order of magnitude smaller and have significantly lower activity per surface area than heterogeneous catalysts. Therefore, the question arises whether or not enzymes are self-phoretic, *i.e.*, able to swim.

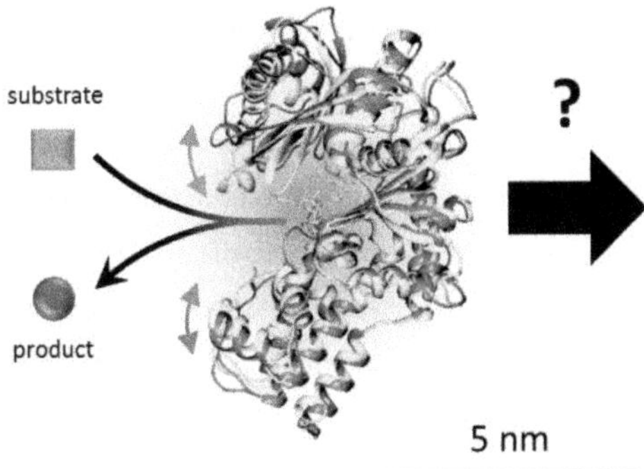

Figure 3.2. Active enzyme diffusion hypothesis. Overlay of two crystal structures of the enzyme hexokinase with (yellow)[47] and without (green)[48] bound substrate. The conformational change (indicated by green arrows) is one of the largest observed for enzymes. The catalytic conversion releases energy (red) at the active site of the enzyme. The direction and velocity of propulsion – if it can be observed – would depend on the details of the underlying mechanism.

Enzymes are already employed by nature in form of motor proteins to accomplish various types of motion. Examples include linear motion by myosin in muscles or kinesin in mitosis[49] and rotational motion of the flagella motor or ATP synthase.[50] It has to be noted that this motion of motor proteins is inherently different from the self-phoresis of enzymes studied here, since motor proteins always actively move relative to a bigger structure (protein fiber or lipid layer), which can be seen as stationary, whereas the research on self-phoresis or enhanced diffusion is studied away from surfaces. This is why the active diffusion of only the F_1-part of F-ATP synthase will be discussed in Chapter 7 and not the complete enzyme embedded in a membrane.

The hypothesis that enzymes are active matter and show enhanced diffusion when they are catalytically active has been proposed in 2010,[7] even though similar observations with different interpretations have been made earlier.[11] FCS measurements were central in the reported observations of active enzyme diffusion. These studies mainly focused on commercially available enzymes such as urease,[7,8,10] catalase,[8,23,51] alkaline phosphatase,[8] and triose phosphate isomerase.[8] To date, urease is the most studied enzyme in the context of active enzyme diffusion. The extent of diffusion enhancement is usually defined as $\Delta D/D_0 = (D - D_0)/D_0$, where D is the diffusion coefficient in presence of the substrate (or another molecule) and D_0 is the diffusion coefficient without reaction (or interaction with another molecule). Initial results ranged from diffusion enhancements $\Delta D/D_0 = 30\,\%$ for urease[7] to 80 % for alkaline phosphatase.[8] In addition to FCS, which is a self-diffusion measurement technique at low sample concentrations, chemotactic measurements were introduced to observe mutual diffusion of enzymes and their substrates as well as enzymes in cascades.[12,24,25] With the growth of the field other enzymes such as fructose bisphosphate aldolase,[9,12] acetylcholinesterase,[10] and hexokinase[12] were studied. A complete list of the enzymes studied in the context of enhanced self-diffusion can be found in Chapter 11. Fructose bisphosphate aldolase is of special interest, since it catalyzes an endothermic reaction and was also reported to show enhanced diffusion in presence of its inhibitor.[9] This report was significant, since it was previously believed that exothermicity is necessary for enzymes to show self-propulsion. With the use of stimulated emission depletion (STED) FCS smaller confocal volumes allowed experimenters to distinguish fast and slow diffusion processes in the same sample, which was interpreted by these authors as "leaps" that an enzyme makes, when it is active.[10]

At present there is no unified explanation for the reported observations of enhanced enzyme propulsion and a number of different theories have been proposed. The first question one can ask is whether there is enough energy released during catalysis to power enhanced diffusion to the extent it has been reported in the aforementioned studies. Since most of these reactions are exergonic, the released energy could, in principle, be channeled into motion to overcome viscous drag. Theoreticians come to different conclusions whether the energy released during the reaction is sufficient. Tlusty calculated that the required energy for urease enhanced diffusion is roughly 25 kJ/mol,[52] which is close to the available 20 kJ/mol of free reaction enthalpy (calculated from the equilibrium constant of the hydrolysis of urea).[53]

$$\mathrm{H_2N\text{-}CO\text{-}NH_2} + \mathrm{H_2O} \longrightarrow 2\,\mathrm{NH_3} + \mathrm{CO_2} \qquad \Delta G_R^{\ominus} \approx 20\ \mathrm{kJ/mol} \tag{3.4}$$

Feng and Gilson came to a different conclusion with a minimal requirement of 100 kJ/mol for urease, which would rule out that the energy of the catalysis would be sufficient for the reported enhanced enzyme diffusion.[14]

Self-diffusiophoresis might also be a possible explanation for enhanced enzyme diffusion. A diffusion enhancement of $\Delta D/D_0 \approx 30\,\%$ was observed for urease,[7] which has a theoretical $D_0 \approx 4.4\ 10^{-11}\ \mathrm{m^2/s}$ and $D_{\mathrm{rot}} \approx 1.3\ 10^{6}\ \mathrm{s^{-1}}$ (for $R_{\mathrm{h}} = 5$ nm, $\eta = 10^{-3}$ Pa s, and $T = 300$ K). Solving Eq. (3.3) for v yields a required propulsion speed of 10 mm/s. This would amount to a travelled distance of 1 µm per catalytic cycle assuming a catalytic rate of $k_{\mathrm{cat}} = 10^{4}\ \mathrm{s^{-1}}$, which is completely unrealistic for a 10 nm enzyme. This rules out self-diffusiophoresis as a possible propulsion mechanism. The theoretical challenge is – if the experimental results are correct – whether there is a mechanism that could explain how the reaction energy is channeled more efficiently into translational motion. Motor proteins reach astonishing efficiencies of 80 %,[54] but move relative to templates. Freely diffusion enzymes would need to direct the energy into directed motion even though they are subject to fast rotational diffusion. Urease, for example, rotates on average 90° in 315 ns in aqueous solution at room temperature.

In the following, novel propulsion theories for enzymes will be discussed, which aim to explain the experimentally observed diffusion enhancement (Fig. 3.3). One proposed explanation is a chemoacoustic effect in which the reaction in the active site of the enzyme creates a pressure wave due to exothermic heat release. Since the active site is usually off center, this pressure wave should propel an enzyme forward by a recoil mechanism.[8] It was later shown that this theory cannot explain the extent of diffusion enhancement observed in experiments and would be a rather small effect ($\Delta D/D_0 \approx 10^{-5}$) due to the large number of degrees of freedom in a protein and damping of the acoustic wave.[28,55]

Another proposed explanation is a temperature increase in the solution due to the heat released from the reaction.[55] The heating depends on the exothermicity of the reaction and the dimensions of the reaction container. If the container is large enough and the enzyme concentration high enough, then thermal conductivity might not be sufficient to transport the reaction heat out of the container. The diffusion increase according to Eq. (2.8) is caused by temperature increase and viscosity decrease, since the viscosity of water is also dependent on temperature. This theory was initially deemed sufficient for the observed diffusion enhancements, but was later questioned by the authors of one of the experimental studies.[56,57]

In the light of enhanced diffusion reports in the absence of heat release (endothermic reactions and interactions with competitive inhibitors and cofactors) a new theory was developed, which was termed fluctuation-induced hydrodynamic coupling. This theory is based on the coupling of the conformational change of the enzyme (due to interactions with substrates, inhibitors, or cofactors) with the surrounding solution. The enzyme's shape change was estimated with a flexible dumbbell model. The shape fluctuations then couple with the surrounding medium and lead to diffusion enhancement.[9,58] Later numerical Brownian dynamics simulations came to the conclusion that fluctuation-induced hydrodynamic coupling does not lead to high enough diffusion enhancements to explain the experimental reports.[59]

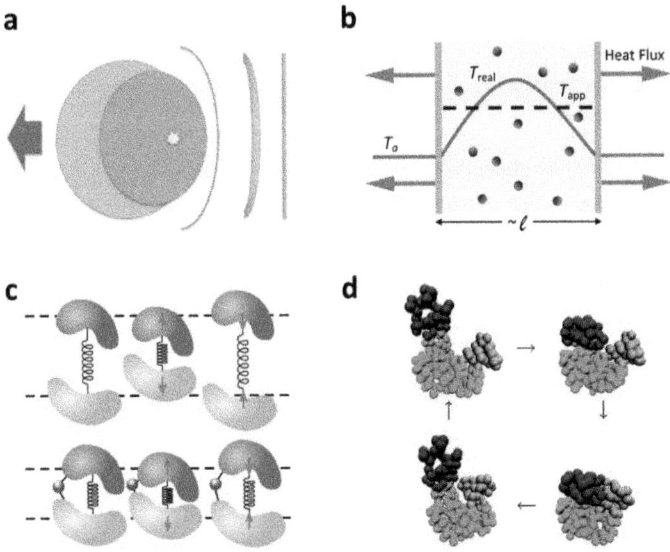

Figure 3.3. Examples for theoretical explanations of enhanced enzyme diffusion. **a** Chemoacoustic propulsion mechanism. The enzyme is propelled by acoustic waves (blue) emitted from the active site (yellow asterisk). Reprinted with permission from Ref. 8. **b** Collective heating. The exothermic reaction leads to heating over the container length scale ℓ. Reprinted with permission from Ref. 55. **c** Fluctuation-induced hydrodynamic coupling. The enzyme, which is approximated as a dumbbell (orange and purple) with a harmonic-like potential (spring), interacts with the substrate/product (blue). Reprinted with permission from Ref. 58. **d** Conformational change. Coarse-grained molecular dynamics simulation of the conformational change of adenylate kinase. Modified and reprinted with permission from Ref. 46.

In an earlier study coarse-grain molecular dynamics simulations were used to simulate the diffusion change of an enzyme undergoing conformational change.[46] The studied enzyme, adenylate kinase, undergoes more than two distinct conformational stages in its catalytic cycle,[60] which would allow for self-propulsion at low Re according to Purcell's scallop theorem. The simulation showed a diffusion enhancement of up to 20 % for adenylate kinase, but the enhancement could nearly fully be accounted for by changes in the hydrodynamic radius of the enzyme due to its conformational change during the catalytic cycle (Eq. (2.8)).

Most theoretical approaches come to the conclusion that active enzyme diffusion is possible, but disagree with the initial experimental studies on the extent of the effect. Therefore, highly accurate diffusion measurement techniques are needed to further test the active enzyme diffusion hypothesis and to reveal the origin of the initial experimental observations. The techniques used in this thesis to undertake these diffusion studies will be discussed in Chapter 4. The experimental results on active enzyme diffusion will be reported in the Chapters 7 and 8.

3.3 Molecular Catalysts & Tracer Diffusion

Motion on a molecular level even smaller than proteins has been achieved with artificial molecular machines. These are molecules which can transfer chemical stimuli into motion and for which the Noble Prize in chemistry was awarded in 2016. One particularly interesting example of a molecular machine is the Feringa motor, which is a rotational molecular machine powered by light.[61] Even though the motion, *i.e.*, conformational change, of molecular machines has been proven, there has only been one report of enhanced diffusion of molecular machines.[62] Similarly also molecular catalysts which are not molecular machines have been reported to exhibit enhanced molecular diffusion.[63] This diffusion enhancement has been proposed to be transferred even to surrounding molecules.[64] In what follows, reports on enhanced molecular diffusion and transfer of motion to tracers will be discussed.

The most studied system in the context of enhanced molecular diffusion are Grubbs catalysts.[13,63,64] Grubbs catalysts contain complexed ruthenium ions and catalyze metathesis reactions of alkenes. During Grubbs metathesis, enhanced diffusion of the catalyst and the solvent have been reported for diffusion NMR measurements (see Chapter 4.2). A nearly 5-

fold diffusion increase of the catalyst and a nearly 2-fold diffusion increase for tracer molecules was initially explained with momentum transfer from the catalyst to the surrounding molecules.[64] However, it was later shown that the apparent increase in diffusion was actually caused by convection artefacts due to heat development in the NMR sample tube.[65] This convection artefact, which will also be discussed later, seems to completely account for the observed apparent diffusion enhancement in Ref. 64. Nevertheless, another very recent study reported "boosted" molecular motion involved in several reactions including a different metathesis reaction.[13] In addition to the previous claims for the catalyst and tracer molecules, these authors also claimed to observe diffusion enhancement for the solvent and the reactant molecules. The central reaction of this study is a copper-catalyzed click reaction, during which diffusion enhancements of over 60 % were measured. The authors of Ref. 13 stated that they took care to avoid convection artefacts with advanced diffusion NMR pulse sequences and control experiments. Some reactions studied in Ref. 13 did not yield enhanced diffusion, which was explained with too low Gibbs free energy release rates during these reactions. The diffusion enhancement during the remaining reactions, including the click reaction, was explained by the authors of Ref. 13 with rapid electronic structure changes in the molecules during the reaction.[13] Within the work of this thesis it was possible to find a different explanation for the reported apparent diffusion enhancement, which is presented in Chapter 10.

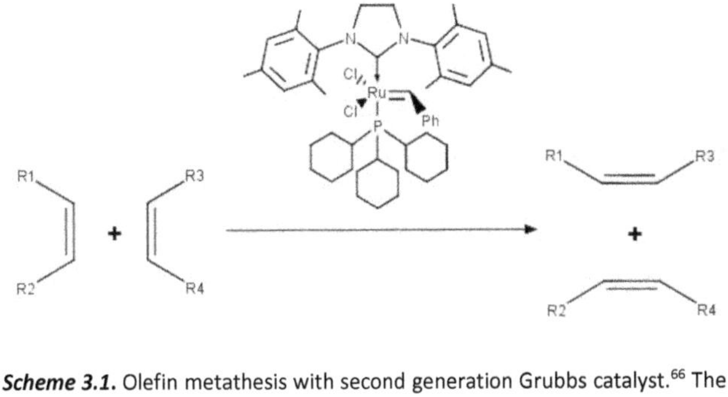

Scheme 3.1. Olefin metathesis with second generation Grubbs catalyst.[66] The residual groups (R1 to R4) are reorganized. Ph corresponds to a phenyl residue.

It is worth mentioning that enhanced tracer diffusion has not only been observed in the presence of active molecular catalysts, but also in the case of active enzymes. The tracers in enzyme studies involved small fluorophores and particles much bigger than the enzyme (up to 2 µm). Several enzymes were tested with the highest diffusion enhancement being on the order of 30-40 %.[67] The study was performed with FCS and DLS in homogenous solutions, *i.e.*, convection effects should be excluded. These results were tested within this thesis with NMR as an absolute and model-free method with a different conclusion (Chapter 8).

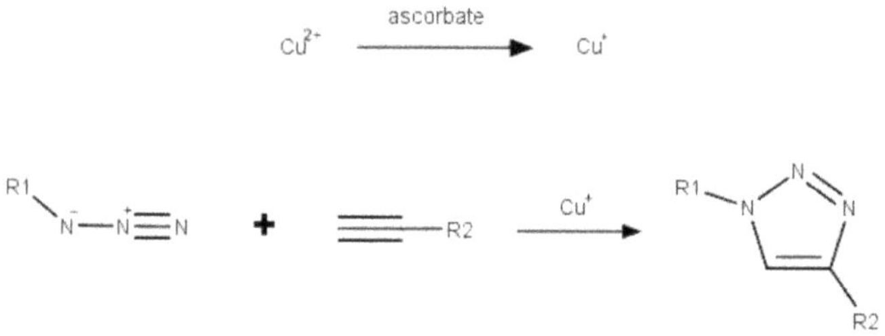

$$Cu^{2+} \xrightarrow{\text{ascorbate}} Cu^{+}$$

$$R1{-}\overset{-}{N}{-}\overset{+}{N}{\equiv}N \; + \; HC{\equiv}C{-}R2 \xrightarrow{Cu^{+}} \text{1-R1-4-R2-1,2,3-triazole}$$

Scheme 3.2. Copper(I)-catalyzed 1,3-cycloaddition.[68] Cu^{2+} is reduced to Cu^{+} (here, by ascorbate), which catalyzes the 1,3-cycloaddition of an azides and an alkyne.

The majority of this thesis carefully examines diffusion of free enzymes in solution, and finds that previous experimental reports of enhanced enzyme diffusion are due to erroneous interpretations of data and artefacts in measurements. There is, however, one effect that shows how chemically-active enzymes can cause hydrodynamic flows via convection. Convection can also be caused intentionally by immobilizing and localizing enzymes near an interface, which can be used to realize an enzymatic micropumps. This work will be discussed in Chapters 6 and 9.

The examples above show that enhanced diffusion is not only a phenomenon hypothesized for enzymes, but also for much smaller, synthetic catalysts. The validity and interpretation of the studies presented above is highly debated and consequently precise and advanced diffusion measurements are needed to cautiously test these hypotheses. Therefore, FCS, diffusion NMR, and other diffusion measurement techniques will be presented in the following chapter.

4 Diffusion Measurement Techniques

"The reputed difficulty of diffusion measurements stems from inherent masochists, like me, who make many of the experiments. We are never satisfied. When we attain coefficients accurate to 10%, we want 2%; when we achieve 2%, we want 0.5%." "This suggests that measurements of diffusion are a Holy Grail requiring noble knights who dedicate their lives to the quest."[69]

Edward L. Cussler

4.1 Fluorescence Correlation Spectroscopy

4.1.1 Single Molecule Fluorescence

Fluorescence studies are important for biochemistry and biophysics due to their high signal to noise ratio and due to the non-invasiveness of the measurement. Improvements in equipment sensitivity and fluorophore stability in the last century led to fascinating measurements on individual fluorescent molecules. Single molecule measurements are technically challenging, but offer insight in dynamic processes, which are inaccessible by ensemble measurement techniques, *e.g.* the fast photophysical dynamics of the yellow fluorescence protein[70] or the rotational subunit motion within active F_oF_1-ATP synthase.[71,72] Even for the most stable fluorescent dyes, which emit 10^6 photons prior to bleaching,[73] it is important to be aware of other photophysical processes in addition to fluorescence. These processes occur to a different extent for different fluorescent systems, but are never completely absent.

It is best to discuss fluorescence with the help of the Jablonski diagram (Fig. 4.1), which depicts the transitions of the fluorophore between several electronically excited states.[73] At room temperature the molecule's ground state is the lowest vibrational state of the S_0 singlet state. Illumination with the resonant wavelength can then excite one electron into one of the higher vibrational modes of the S_1 state. After very fast vibrational relaxation (typically 10^{-12} s) to the lowest vibrational state of S_1, several decay pathways are possible. Fluorescence is the most interesting one for detection techniques. Fluorescence relaxes the electron to the ground state S_0, which emits a photon of longer wavelength, *i.e.*, lower energy, than the excitation photon. This stochastic process is described with a characteristic rate k_{10}. Here, k_{10} is defined to also include non-radiative relaxation processes from S_1 to S_0. This includes collisional quenching, where the excited fluorophore transfers the energy to a quencher molecule Q. The excited quencher molecule Q* then relaxes the energy to the surrounding. Hence, k_{10} is also dependent on the quencher concentration.

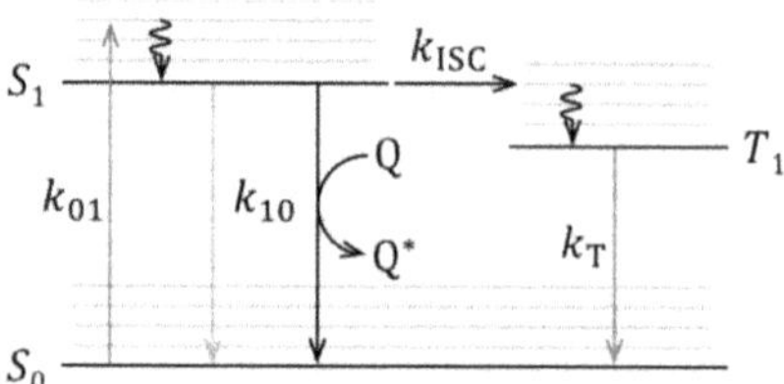

Figure 4.1. Jablonski diagram. Photophysical processes: Excitation (blue), fluorescence (green), and phosphorescence (red). Non-radiative processes in black: intersystem crossing (ISC), collisional quenching with quencher molecule Q, and vibrational relaxation (internal conversion, wavy arrows).

Another photophysical pathway is intersystem crossing (ISC), where the fluorophore is changing from a singlet S_1 state into an excited triplet T_1 state. Since this requires an electron spin flip, which is symmetry forbidden, it is slow compared to the other transitions. Its rate is described by k_{ISC}. Other molecules or ions in the solution surrounding the fluorophore can drastically change k_{ISC} due to spin-orbit coupling. Phosphorescence, which is the emission of a photon during the transition from the T_1 to the S_0 state, is also symmetry forbidden and therefore has a low rate (k_T).

The fluorescence lifetime τ_{F} is a measure of how long the fluorophore is on average in the excited state. It is dependent on k_{10} and k_{ISC}:[73]

$$\tau_{\mathrm{F}} = \frac{1}{k_{10} + k_{\mathrm{ISC}}} \tag{4.1}$$

τ_{F} is usually on the nanosecond timescale, if fluorescence is dominant. When phosphorescence is dominant, then the lifetime is on the order of microseconds to seconds. τ_{F} can be determined with time-correlated single photon counting (TCSPC) experiments, which are useful, since the rate constants themselves are difficult to access. The fluorescence lifetime can also give information about the surrounding of the fluorophore, since it is dependent on the quencher concentration. This will be examined and discussed in greater detail in this thesis in Chapter 7.

4.1.2 Fluorescence Detection from a Confocal volume

Fluorescence correlation spectroscopy (FCS)[73,74] uses – like other fluorescence microscopy techniques – a confocal geometry for the detection of single molecules (Fig. 4.2). The optical setup includes a laser excitation beam, which is focused by an objective into the sample. This leads to an approximated 3D Gaussian intensity profile of the excitation light. For detection, the excitation light is collected by the same objective and spectrally filtered by a dichroic mirror. The emission light is passed through a pinhole. This narrows the detection volume of the emission to a small volume near the focal plane. The combination of excitation and emission optics define the confocal volume, which is usually on the order of femtoliters. The confocal volume has the shape of a (prolate) spheroid with the radii ω_0 in the radial direction and z_0 in the axial direction. It should be noted that the confocal volume does not have a sharply defined surface, but rather decays towards its outside. Its surface is therefore defined as the area, where the intensity has decayed by a factor of e^{-2}.[73]

Since single photons have to be detected, avalanche photo diodes (APDs) are usually used for signal amplification. Multidetector FCS, fluorescence cross correlation spectroscopy (FCCS), allows more complex detection schemes with multiple APDs, which collect the light separated by beam splitters. It is possible to further separate the emitted light according to its polarization (to measure rotational diffusion), by wavelength (to measure Förster resonance energy transfer), or split the intensity evenly (to average out uncorrelated detector noise).[74] Additionally, it is possible to observe the interaction of two fluorescent species at different wavelengths to determine their association or dissociation via fluorescence cross correlation.[75]

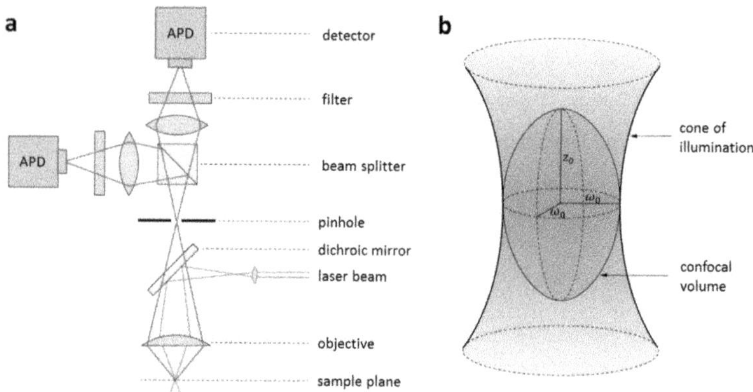

Figure 4.2. FCS instrumentation and optics. **a** Layout of a two detector FCS setup. Excitation path in green and emission path in red. **b** Confocal volume dimensions. Figures recreated and modified from Ref. 73.

4.1.3 Autocorrelation Function and Fitting

For FCS measurements the concentration of fluorophores is generally low enough so that on average only one single fluorophore is present in the confocal volume. The fluorophores are constantly diffusing in and out of the confocal volume, which leads to fluorescence intensity fluctuations/photon bursts over the observation time. The timescale of these fluctuations is therefore dependent on the translational diffusion coefficient of the fluorophore (or the sample labeled with the fluorophore). The autocorrelation function (ACF) G is used to describe these intensity I fluctuations over time t as a function of the lag time τ between two signals.[73]

$$G(\tau) = \frac{\langle I(t) \cdot I(t+\tau) \rangle}{\langle I(t) \rangle^2} \tag{4.2}$$

To yield data of high enough quality for diffusion analysis, intensity time traces of several minutes (ideally 30 min depending on the count rate) have to be collected to average out statistical fluctuations and noise. Drifts in intensity, fluorescing impurities or aggregates recorded over this period can have drastic effects on the quality of the autocorrelation function.

The diffusion coefficient is extracted from the ACF by fitting the following equation,[73] which is only valid if diffusion is the only source of intensity fluctuations, which is an idealized case never reached in practice due to the aforementioned photophysical processes.

$$G(\tau)_{\mathrm{D}} = \frac{1}{N}\left(\frac{1}{1+\tau/\tau_{\mathrm{D}}}\right)\left(\frac{1}{1+(\omega_0/z_0)^2\,\tau/\tau_{\mathrm{D}}}\right)^{\frac{1}{2}} \tag{4.3}$$

Here, τ_{D} is the translational diffusion time. At $G(0)$ the ACF is equal to $1/N$, where N is the average number of fluorophores in the confocal volume. ω_0 and z_0 have to be determined by a calibration experiment using a fluorophore with a known diffusion coefficient. The translational diffusion coefficient can then be determined with the help of the following equation:

$$D = \frac{{\omega_0}^2}{4\,\tau_{\mathrm{D}}} \tag{4.4}$$

At very low fluorophore concentrations, which are typical for FCS, D represents the self-diffusion coefficient. At higher concentrations, FCS determines the mutual diffusion coefficient, since it is sensitive to the relaxation of concentration fluctuations.[76] As mentioned, several other (photo-)physical effects influence the ACF (Fig. 4.3).[77] Translational diffusion is usually the slowest of these effects. The fastest effect is antibunching, which is caused by the fluorescence lifetime, since the fluorophore needs time to relax to the ground state to be ready to start the next excitation-emission-cycle. Rotational diffusion appears in the ACF on the order of nanoseconds to microseconds, but drastically varies with the size of the fluorophore/sample. Triplet lifetime is also very important for FCS analysis and should usually be included in the fitting of the ACF, since the ramifications of triplet lifetimes can be on the same timescale as diffusion times (see also Chapter 7).

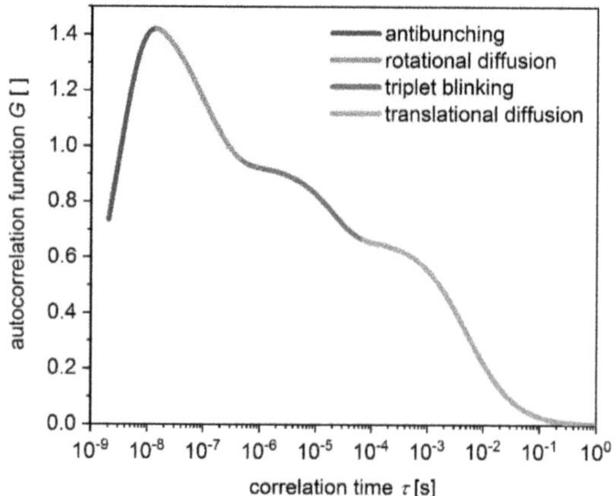

Figure 4.3. Timescales for various correlated processes in FCS. Values for simulated ACF: $\tau_F = 3$ ns, rotational correlation time $\tau_{rot} =$ 50 ns, $\tau_T = 20$ µs, and $\tau_D = 5$ ms.

Fluctuations due to triplet state transitions contribute to the ACF via multiplication of the triplet term in the ACF G_T:[73]

$$G(\tau) = G(\tau)_D \, G(\tau)_T \tag{4.5}$$

The effect of the triplet state dynamics on FCS can be visualized as blinking of the fluorophore, while diffusing through the confocal volume. This leads to dark intervals within the photon bursts and moves the signal to lower correlation times in the ACF. G_T is a function of the triplet relaxation time τ_T and the fraction of molecules in the triplet state T:[73]

$$G(\tau)_T = 1 + \frac{T}{(1-T)} \exp\left(-\frac{\tau}{\tau_T}\right) \tag{4.6}$$

The rate constants (Fig. 4.1) of the photophysical effects discussed above are also dependent on the concentration of quencher molecules in solution. This can significantly alter the ACF and can, if not correctly accounted for, lead to misinterpretations of FCS data, which will be discussed in Chapter 7.

4.2 Nuclear Magnetic Resonance Spectroscopy

4.2.1 Nuclear Magnetization

Additionally to fluorescence, which is a property of the electronic structure of a molecule, nuclear properties can also be used for diffusion measurements. The magnetic moment $\boldsymbol{\mu}$ of an atomic nucleus is connected to its angular momentum $\boldsymbol{I}$ via the gyromagnetic ratio γ:

$$\boldsymbol{\mu} = \gamma \boldsymbol{I} \qquad \text{with} \qquad I_z = \hbar m_I \tag{4.7}$$

The z-component of the angular momentum I_z is quantized with the magnetic quantum number m_I. The z-component of magnetic moment μ_z is therefore also quantized and depends on m_I and the spin quantum number I:

$$\mu_z = m_I \gamma \hbar \qquad \text{with} \qquad m_I = I, I-1, I-2, \ldots, -I \tag{4.8}$$

For protons only two nuclear spin states exist, since $I(^1\text{H}) = 1/2$. These states are degenerate in absence of an external magnetic field. In a static external magnetic field $\boldsymbol{B}_0$, the magnetization's energy is changed depending on their orientation relative to $\boldsymbol{B}_0$.

$$E = -m_I \gamma \hbar B_0 \tag{4.9}$$

$$\Rightarrow \Delta E = \gamma \hbar B_0 \qquad \text{for} \qquad \Delta m = 1 \tag{4.10}$$

This energy difference is used in nuclear magnetic resonance (NMR) spectroscopy to resonantly excite transitions.[78] Using $\Delta E = h\nu_0$, the Larmor frequency $\nu_0 = \gamma B_0$ can be determined. ν_0 is typically in the range of hundreds of MHz for modern NMR spectroscopy in the strong B_0 fields of superconducting magnets. Local changes in the magnetic field caused by the electronic structure of molecules can be detected as frequency changes in NMR, which makes NMR a powerful tool for structure determination and other applications in chemistry. Differences in the Larmor frequency are expressed with the dimensionless quantity δ, which is called the chemical shift. δ is expressed relative to a standard sample (*e.g.* tetramethyl silane for ^{1}H spectra):

$$\delta = \frac{\nu_{\mathrm{sample}} - \nu_{\mathrm{standard}}}{\nu_{\mathrm{standard}}} \tag{4.11}$$

The macroscopic magnetization $\boldsymbol{M}$ is the sum of all k magnetic moments in an NMR sample:

$$\boldsymbol{M} = \sum \boldsymbol{\mu}_k \tag{4.12}$$

The equilibrium magnetization $\boldsymbol{M}_0$ points along $\boldsymbol{B}_0$. $\boldsymbol{M}_0$ is relatively small, since only a small fraction of magnetic moments aligns with $\boldsymbol{B}_0$. This is due to the Boltzmann distribution of moments over the small energy gap ΔE.

When an oscillating magnetic field $\boldsymbol{B}_1$ is applied perpendicular to $\boldsymbol{B}_0$ with the Larmor frequency, $\boldsymbol{M}$ is (partially) deflected into the x-y plane and starts precessing around z. Since the frequency of $\boldsymbol{B}_1$ is in the radio frequency spectrum, it is often referred to as RF pulse. The angle of deflection α is proportional to the strength B_1 and length t_{p} of the RF pulse:

$$\alpha = \gamma B_1 t_{\mathrm{p}} \tag{4.13}$$

The precession of $\boldsymbol{M}$ after an RF pulse can be detected as a so-called free induction decay (FID) with receiver coils. The decay of the FID over time due to dephasing of the magnetic moments is dependent on the spin-spin relaxation constant T_2^*.

Relaxation processes play an important role in NMR, since they influence the experiment repetition rate, the line intensity, the line broadness, and can in general be used to probe properties of the sample. T_1 is the spin-lattice relaxation constant, which describes the time with which the magnetization $\boldsymbol{M}_0$ recovers. T_1 is important for the repetition rate at which NMR experiments can be performed, since longitudinal magnetization (z-component of $\boldsymbol{M}$) has to reestablish before a new pulse sequence can be started. T_1 is dependent on the interaction of the nuclear magnetization with surrounding fluctuating magnetic moments.

These are *e.g.* caused by thermal motion of other nuclear magnetic moments or much stronger magnetic moments of paramagnetic species. In the Bloembergen-Purcell-Pound (BPP) theory,[79] these interactions are described by the correlation time τ_c. T_1 attains a minimum for a certain τ_c, where the spin precession is in resonance with the fluctuating magnetic field in its surrounding (Fig. 4.4).

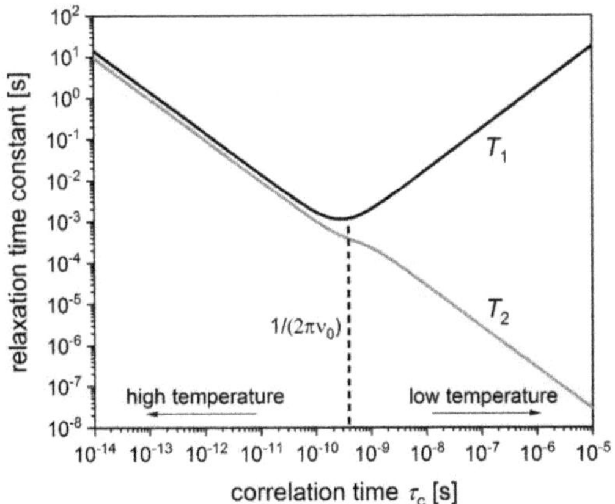

Figure 4.4. Dependency of relaxation time constants on the correlation time. Exemplary T_1 and T_2 values were calculated with the Bloembergen-Purcell-Pound (BPP) theory for ^{1}H-^{1}H interaction.[79,80]

T_2^* describes the dephasing of the precessing magnetic moments $\boldsymbol{\mu}_k$ and therefore influences the transverse component of $\boldsymbol{M}$. The dephasing is caused by spin-spin interactions and leads to variations of the Larmor frequencies, which in turn leads to line-width broadening. T_2^* can be written as:

$$\frac{1}{T_2^*} = \frac{1}{T_2} + \frac{1}{T_{2,\mathrm{ih}}} \tag{4.14}$$

where T_2 is the spin-spin relaxation constant in the homogeneous and $T_{2,\mathrm{ih}}$ in the inhomogeneous magnetic field. $T_{2,\mathrm{ih}}$ highly depends on the experimental conditions, *i.e.*, the quality of the magnetic field. T_2 can be calculated from BPP theory[79] and monotonically decreases with increasing τ_c (Fig. 4.4). The influence of relaxation times on diffusion NMR will be covered in Chapter 10.

Since some of the NMR samples of this thesis are proteins (see Chapter 8), the important features of proteins in NMR will now be discussed. Wild-type – non-enriched and non-labeled – proteins contain several nuclear species, which are interesting for NMR spectroscopy: ^{1}H, ^{13}C and ^{17}O are abundant in side chain and back bone, whereas ^{15}N only occurs in the backbone. For structure determination the probed isotope is important, but for diffusion measurements only the sensitivity of NMR towards the signal of a certain isotope is important under the assumption that the diffusive motion of all parts of the protein is the same. The sensitivity of NMR spectroscopy towards certain isotopes of an element depends on the spin quantum number I, the natural abundance and the magnetogyric ratio γ of the isotope. Usually sensitivity is compared relative to ^{1}H. ^{13}C, ^{15}N and ^{17}O have relative sensitivities of 0.016, 0.001 and 0.029, respectively. ^{1}H is the ideal candidate for protein diffusion NMR, since H is the most abundant element in proteins and ^{1}H is the most sensitive isotope in proteins.

4.2.2 Pulsed Field Gradients

When NMR spectroscopy is used for diffusion measurements it is often called diffusion-ordered spectroscopy (DOSY),[81] pulsed field gradient NMR (PFG-NMR),[82] or pulsed gradient spin echo NMR (PGSE NMR) spectroscopy.[22] All methods have in common that they use magnetic field gradients to measure diffusion. This is achieved by altering the static magnetic field B_0 with a smaller gradient G usually in the z-direction, which leads to a spatially varying total magnetic field strength B'.

$$B'(z) = B_0 + z\frac{dB}{dz} = B_0 + zG \tag{4.15}$$

As B' is a function of z, so is the Larmor frequency ν_0'. This 'labels' the nuclei with a Larmor frequency unique to their position in z. Due to translational diffusion the nuclei move also along z to a new position with a different ν_0'. Stimulated[83] and spin echo experiments[84] can use this difference to quantify diffusion. In the following, the focus will be on the stimulated echo as an example.

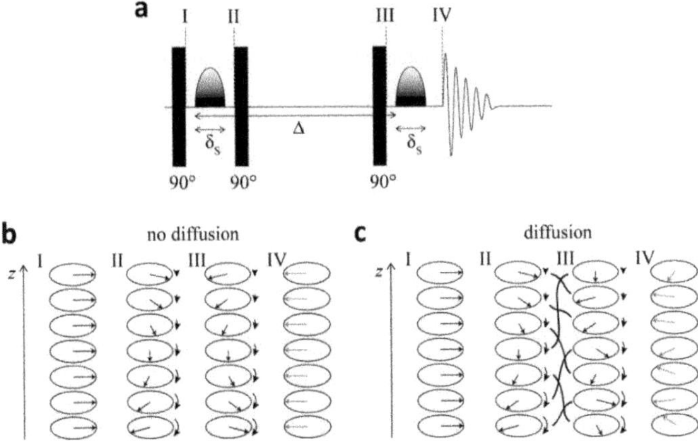

Figure 4.5. Stimulated echo diffusion NMR experiment. **a** Stimulated echo pulse sequence with sinusoidal gradient pulses $\delta_S = \frac{\pi}{2}\delta_G$ and diffusion time Δ. **b** and **c** Idealized precession of magnetic spins at different positions without (**b**) and with (**c**) diffusion. The positions I to IV are marked in the pulse sequence in **a**. Reprinted with permission from Ref. 2 (Chapter 8).

In the stimulated echo PFG-NMR experiment (Fig. 4.5), the magnetization $\boldsymbol{M}$ is deflected into the x-y plane with a 90° pulse. Then a gradient pulse of strength G and length δ_G is applied to ensure that the nuclei have a Larmor frequency that depends on their position within the sample. Directly afterwards the magnetization is deflected to the $-z$ direction with another 90° pulse. This is advantageous, when T_2 is significantly shorter than T_1, since there is very little time for dephasing of the magnetic moments due to T_2 and the intensity loss during the diffusion time Δ is only due to T_1. The nuclei are changing position during Δ due to translational diffusion and are then deflected back into the x-y plane by another 90° pulse.

The combined 180° pulse changed the order of the dephased magnetic moments (Fig. 4.5, step II to III). The magnetic moments can now rephase during a second gradient pulse. If the nuclei would hypothetically stay in the same place and if there were no diffusion, then they would now all rephase, since they all have the same Larmor frequency during the first and second gradient pulse. If they are displaced from their original positon (*e.g.* by diffusion) the intensity loss is a measure of their diffusion coefficient.

The signal attenuation due to diffusion during a PFG-NMR experiment is described by the Stejskal-Tanner equation:[84]

$$I(G) = I_0 \cdot \exp(-b(G) \cdot D) \tag{4.16}$$

Here, I is the signal intensity at the end of the pulse sequence, I_0 the intensity in the absence of a gradient G, b the diffusion weighting factor, which is a function of several experimental parameters, and D the translational diffusion coefficient. Eq. (4.16) can either be plotted linearly as $\frac{I(G)}{I_0}$ (Fig. 4.6) or in a semi-log plot as $\ln\left(\frac{I(b(G))}{I_0}\right)$ (Fig. 4.8). For gradients of sinusoidal shape, b has the form of[22]

$$b(G) = \gamma^2 G^2 \left[\delta_G{}^2 \left(\Delta - \frac{\pi}{8}\delta_G\right)\right] \tag{4.17}$$

In contrast to FCS, PFG-NMR does not require labeling of the sample. Therefore the unaltered sample is observed. Additionally, multiple species can be observed with one PFG-NMR experiment if they consist of the same isotope and have spectrally resolved chemical shifts. This also allows one to simultaneously measure the diffusion of the solvent or other tracers, which can be used to check for effects such as viscosity changes and heating of the sample. Since PFG-NMR directly measures the displacement in z, no additional model is needed to obtain diffusion constants. In contrast, in FCS one requires a reference measurement to determine the focal volume. On the other hand, the single molecule FCS method is much more sensitive and needs roughly 10^6 times less sample then NMR methods. This is due to the weak signals in NMR experiments.

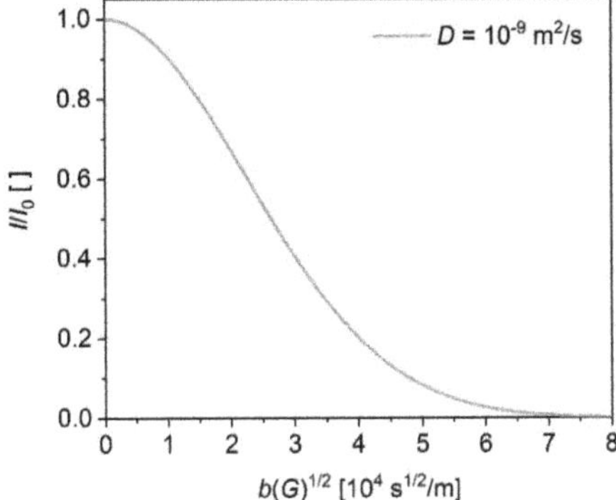

Figure 4.6. Signal attenuation in diffusion NMR experiment according to the Stejskal-Tanner equation.

4.2.3 Advanced Pulse Sequences

Advanced PFG-NMR experiments can be used to correct for effects such as convection or concentration changes, which would otherwise be erroneously interpreted as contributions to diffusion. Thermal gradients within a sample tube can lead to convection for solvents with low viscosity. Even small convective currents can mask diffusive motion and lead to significant signal attenuation. Convection is usually corrected with Pulsed Gradient Double Stimulated Echo (PGDSTE) sequences, which use two consecutive stimulated echo experiments to correct for flow with constant velocity, *i.e.*, convection, during the NMR experiment.[85]

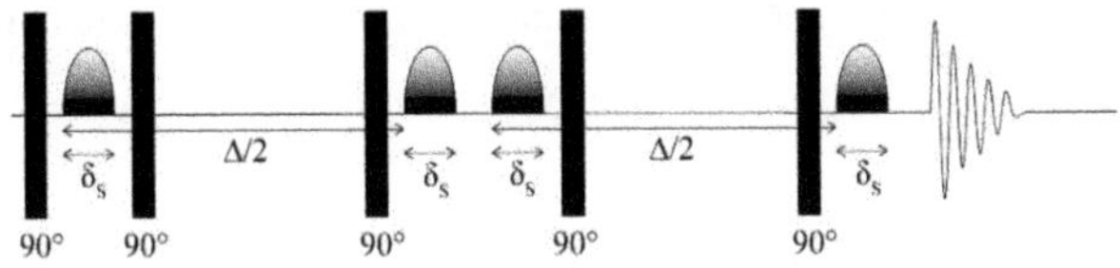

Figure 4.7. Pulsed gradient double stimulated echo sequence for convection suppression. The diffusion time Δ is split into two parts.

Another problem with PFG-NMR measurements is their sensitivity towards intensity changes of the observed species. If the signal intensity of a peak is changing over time due to *e.g.* changes in concentration or in the spin relaxation, then this is hard to distinguish from diffusion changes, since both are determined from the same intensity data. This artefact is particularly problematic, when the relative intensity increase or decrease is strong over the timescale of the PFG-NMR experiment. Since the gradients usually are pulsed in monotonically increasing order, diffusion attenuation of the signal and signal changes due to other effects are both functions of time, which is under certain circumstances not distinguishable in the fit and can misleadingly still lead to very good fit (R^2) values, but incorrect diffusion data. A constant intensity change over time can be expressed as

$$I(t) = I_0(t) \cdot \exp\big(-b\big(G(t)\big) \cdot D\big) \quad \text{with} \quad I_0(t) = I_0(0) + kt. \tag{4.18}$$

Here, the intensity is changing with a constant rate k over time t. $I(t)$ is also dependent on the gradient order $G(t)$. Different influences on the gradient order have been simulated for a negative rate k in Fig. 4.8. To some degree intensity changes can be corrected for by using permutated, *i.e.* random, gradient strengths.[86] This way the intensity change due to artefacts is spread out over multiple gradients. Fits might have poorer quality, but usually include the real diffusion coefficient within the confidence interval. The simulated experiment with random gradient order (green), leads to the same slope and therefore the same diffusion coefficient as the unperturbed measurement (black). Increasing and decreasing orders lead to diffusion coefficients that are respectively too high or too low. An addition to the random gradient method is the use of a sliding window for fitting the diffusion data over random gradient orders. This allows simultaneous high time resolution and corrections to a changing intensity.[86]

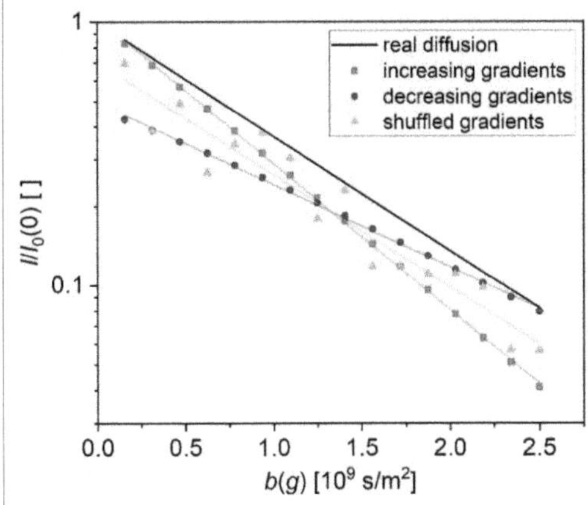

Figure 4.8. Influence of gradient order on diffusion NMR experiments with additional intensity change. The data was simulated assuming a constant intensity loss over the course of the diffusion experiment. Colored lines indicate linear regressions.

4.3 Other Techniques

Besides the methods presented here in detail there are also several other diffusion measurement techniques which should be mentioned in the context of (protein) diffusion without claiming completeness.

4.3.1 Other Correlation Methods

Besides FCS also other methods are based on correlation approaches. In dynamic light scattering (DLS),[87] the scattering signal of particles or macromolecules in solution are correlated. Since the scattering intensity fluctuates due to interference changes due to Brownian motion, the diffusion coefficient can be extracted from the ACF. DLS is especially suited for larger particles and macromolecules, since the intensity of Rayleigh scattering is proportional to the radius of the scatterer to the power of six. The measurement is usually faster than an FCS or PFG-NMR measurement, but also needs concentrations as high as those in PFG-NMR experiments (micromolar for proteins). An advantage of DLS is that it does not require sample labeling.

4.3.2 Tracking Microscopy

For bigger particles tracking microscopy has been proven to be a versatile tool for the determination of diffusion coefficients. Here, the particle is observed over a sufficiently long time period to directly calculate the mean square displacement from the video data.[88] This approach allows one to directly distinguish between Brownian motion and linear propulsion,[37] which is difficult with the other methods presented here. Tracking microscopy has been extensively used to study the propulsion of Janus microparticles that have enzymes attached to one halve of the particle,[39] but observation of the free diffusion of individual proteins in solutions with aqueous viscosity is very challenging and has not fully been achieved yet. This is due to the low intensity and fast movement of individual proteins, which does only allow short signal averaging. A research group has recently succeeded in tracking individual fluorescently labeled urease molecules in highly viscous polymer solutions with total internal reflection fluorescence (TIRF) microscopy, which allows high contrast fluorescence imaging.[89]

Tracking microscopy can also be used to analyze flow phenomena in systems.[88] This way convection can be identified and quantified, which is of interest for microfluidic systems. Some of these systems can be powered by immobilized enzymes as will be discussed in Chapter 6.

4.3.3 Trapping Methods

Trapping methods can be used to create a potential well, where particles or molecules of interest are trapped. This has been done with optical and magnetic fields for nanoparticles, which lead to breakthroughs in the research on motor proteins.[90,91] These trapping methods require the addition of rather large particles to the protein in order to exert sufficient trapping force. Anti-Brownian electrophoretic (ABEL) traps utilize a different approach.[92,93] Here, the diffusion of the sample is counteracted with a feedback-loop to keep it centered in the microscope's object plane. The sample is detected via fluorescence and moved with electrophoretic forces that either act on the protein or the buffer solution. The electrodes are arranged around the flat sample chamber. The advantage of ABEL traps is that smaller objects can be trapped than in optical or magnetic traps. The trapping of single proteins allows much longer observation times than in FCS. The diffusion coefficient can be calculated from the input and feedback signal that is used to move the particle/solution.

4.3.4 Mutual Diffusion Measurement Techniques

All methods discussed above measure self-diffusion coefficients under the applied conditions. Other methods allow to determine mutual diffusion coefficients. In the recent years flow channel experiments with concentration gradients of several enzymes including substrates have gained popularity.[12,23–25] The gradient can be shaped via microfluidic channels, which use the mixing in a direction perpendicular to the laminar flow to determine diffusion coefficients.[10] Several interesting results were presented claiming chemotaxis,[12] antichemotaxis[10] and co-diffusion of enzymes in a cascade.[25] Although those results are impressive, it might be worth mentioning that differences in chemical potential can drive motion in mutual diffusion experiments, which might not be present in self-diffusion experiments. Therefore chemotactic observations should not be misinterpreted as diffusion enhancement. This is evidenced in observations of enhanced diffusion, which did not require catalytic activity, *e.g.* for dye molecules and passive polymers.[94]

4.3.5 Computational Methods

Hydrodynamic properties of proteins can be estimated using computational methods.[95] A valuable tool for this is the HYDRO software suite,[96,97] which consists of several public domain computer programs, and allows the use of protein data bank (PDB) files as input. These structural data files are mainly from crystallography, NMR, and cryogenic electron microscopy experiments and are used by the HYDRO software to build a rigid shell of evenly sized spheres. This shell is then used to calculate the hydrodynamic properties of the protein. The software HYDROPRO calculates translational and rotational diffusion coefficients and sedimentation properties for ultracentrifugation experiments.[96] The software HYDRONMR calculates correlation times and the ratio of T_1 and T_2 in addition to the diffusion properties.[97] These programs are useful to estimate diffusion coefficients of several oligomeric or conformational states of a protein, if the corresponding structural data is available.

5 Enzymes Properties

"Any living being is a reflection of its enzyme arsenal. We are and do what our enzymes permit."[98]

Christian de Duve

Each enzyme has unique characteristics and functionality, which can directly affect the methods and conditions best suited for diffusion studies on them. The activity of an enzyme dictates how long measurements can be performed until the substrate is consumed. Binding constants between the enzyme's subunits, which can also be a function of the substrate concentration, affect the practicable enzyme concentration range. Additionally, primary to tertiary structure define which amino acid residues are accessible for covalent labeling. These parameters have to be considered to select the appropriate measurement technique and even more importantly the required controls for diffusion measurements to avoid misinterpretation of the data (see Chapter 7). In the following, all enzymes used in this thesis will be discussed in order of increasing molecular weight. All images of protein structures in this chapter were created with the software UCSF Chimera.[99]

5.1 Alkaline Phosphatase

Alkaline phosphatase is part of the enzyme family of phosphatases, which catalyze the hydrolysis of phosphomonoesters. In humans alkaline phosphatase has a medical importance and is related to fatty acid absorption and obesity.[100] A very common application of alkaline phosphatase is in enzyme-linked immunosorbent assays (ELISA), where it is conjugated to antibodies.[101] Experiments in the context of this thesis use mammalian intestinal alkaline phosphatase from calf intestinal mucosa.

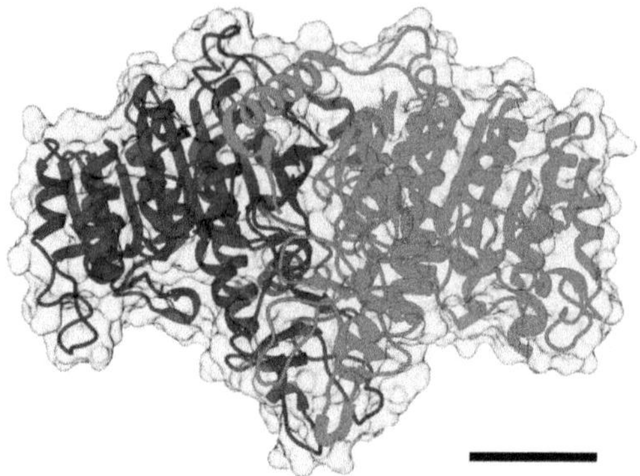

Figure 5.1. Structure of the mammalian intestinal alkaline phosphatase. The x-ray structure of the homodimer from rat intestinal alkaline phosphatase is shown (PDB entry 4KJG).[102] Each monomer is individually colored. Scale bar 2 nm.

Alkaline phosphatase (Fig. 5.1) is a dimeric glycoprotein with a mass of ca. 150 kDa and two zinc ions in each active site.[103] Magnesium ions are essential for stability and activity of alkaline phosphatase, but Mg^{2+} binds to a different site than zinc.[101] The enzyme has many surface-exposed lysines,[102] which can for instance be used for labeling with N-hydroxysuccinimide (NHS) activated dyes for FCS experiments.[8] Native cysteines are not as well suited for labeling of alkaline phosphatase, since most of them are involved in disulfide bridges.[101]

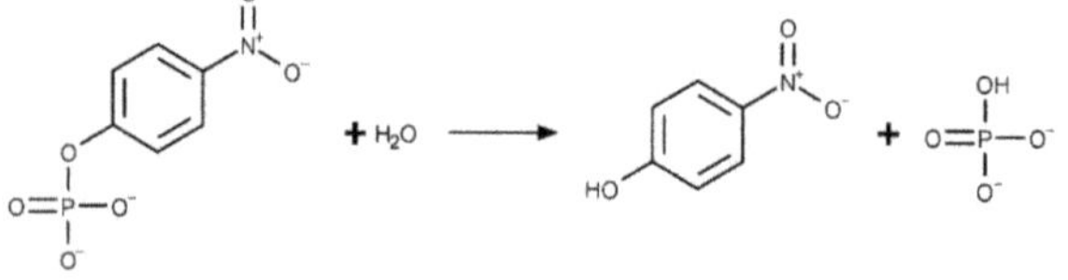

Scheme 5.1. Phosphatase assay reaction. p-Nitrophenylphosphate (NPP) is hydrolyzed to p-nitrophenol, which has an absorption maximum at 405 nm.

Alkaline phosphatase catalyzes phosphorylation and transphosphorylation at pH 8 or higher depending on the used buffer and substrate.[101] Some buffers increase the activity of the enzyme by acting as phosphate-acceptors and therefore allowing transphosphorylation.[101] Under optimal conditions alkaline phosphatase reaches turnover rates of around 15,000 s^{-1}. The product, phosphate, is a competitive inhibitor of alkaline phosphatase,[101] which leads to a relatively fast decay of the activity over time. Activities can be easily observed for the common substrate p-nitrophenolphosphate via the absorption of the product at 405 nm (Scheme 5.1).[101] For other substrates phosphate assays can be used to detect the produced amount of phosphate.[104]

5.2 Fructose-Bisphosphate Aldolase

Fructose-bisphosphate aldolase (from now on aldolase) is a ubiquitous enzyme, which is part of glycolysis and gluconeogenesis.[105] All studies of enhanced enzyme diffusion of aldolase use rabbit muscle aldolase.[9,12,106,107]

Aldolase is a homotetramer with a total mass of 158 kDa (Fig. 5.2).[105] The tetramer, dimer and monomer are in equilibrium with the tetramer being the most abundant oligomer at enzyme concentrations above 10 nM.[106,108] Aldolase dissociates in the presence of its substrate,[109] which makes it difficult to reliably measure diffusion coefficients, when the enzyme is active. The active site of aldolase does not contain any metals and both, the dimer and the monomer, are also active.[110] Aldolase can be labeled with maleimide-activated dyes at its native cysteine residues.[9] Labeling lysine residues is not advisable for aldolase, since its active site contains lysines.[105]

Aldolase catalyzes the reversible cleavage of fructose-1,6-bisphosphate (FBP) to glyceraldehyde 3-phosphate (G3P) and dihydroxyacetone phosphate (DHAP) (Scheme 5.2). The forward reaction in the direction of glycolysis is endothermic. Aldolase catalyzes FBP lysis at a rate of ca. 10 s^{-1}.[105] Known inhibitors of aldolase include phosphate, pyrophosphate, and nucleotides.[105,111] Its activity can be determined in a coupled cascade assay of aldolase, triosephosphate isomerase and α-glycerophosphate dehydrogenase via an optical absorption change at 340 nm due to consumption of NADH by α-glycerophosphate dehydrogenase (see Chapter A.1.2).[112]

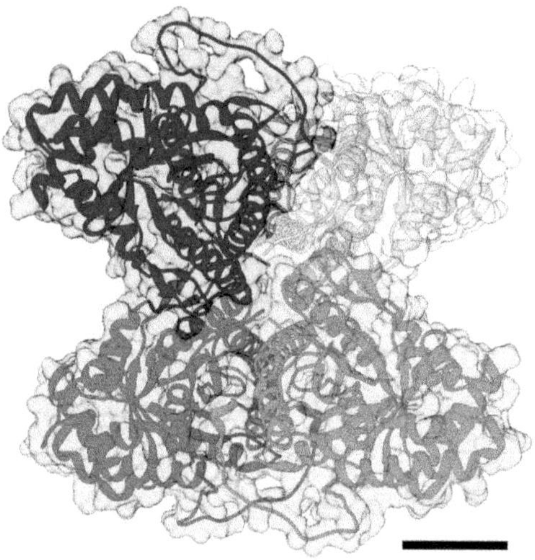

Figure 5.2. Rabbit muscle fructose-1,6-bisphosphate aldolase. Crystal structure of the aldolase homotetramer from PDB entry 1ZAH.[113] Each monomer is individually colored. Scale bar 2 nm.

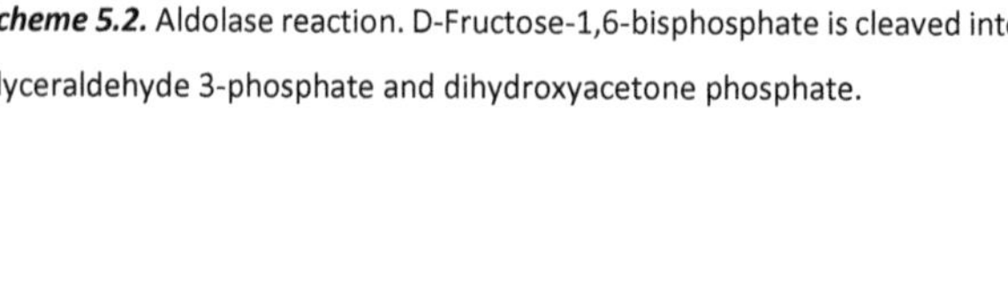

Scheme 5.2. Aldolase reaction. D-Fructose-1,6-bisphosphate is cleaved into glyceraldehyde 3-phosphate and dihydroxyacetone phosphate.

5.3 Urease

Urease is a large, homohexameric enzyme, which can mainly be found in plants, fungi and bacteria. The main source of urease for research purpose is jack bean (*Canavalia ensiformis*).[114] Therefore, other seed protein contaminations can commonly be found in urease products (see Chapter A.2.3).[115] During the hydrolysis of urea, which is catalyzed by urease, ammonia is produced, which leads to pH increase. This pH change is utilized by *Helicobacter pylori* to penetrate gastric mucosa.[116]

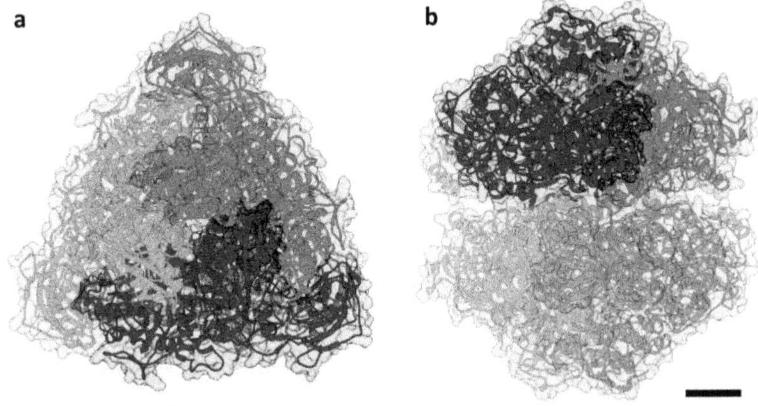

Figure 5.3. Urease from jack bean (*Canavalia ensiformis*). Crystal structure of the urease homohexamer from PDB entry 3LA4.[117] **a** Top view. **b** Side view. Each monomer is individually colored. Scale bar 2 nm.

Urease consists of six identical subunits with a molecular mass of 90 kDa, which each contains two nickel ions in its active site. The monomers form two trimer subunits, which arrange in a hexamer of ca. 10 nm in diameter (Fig. 5.3).[117] Urease can be labeled via its amine and thiol groups, but thiol-reactive labels should not be used in great excess, since the enzyme has a cysteine residue with low maleimide reactivity, which is essential for its catalytic function.[118]

Maleimides,[7] N-succinimidyl(4-iodoacetyl)aminobenzoate (SIAB),[119] glutaraldehyde[120] and NHS-esters[121] have *e.g.* been used for urease immobilization and labeling.

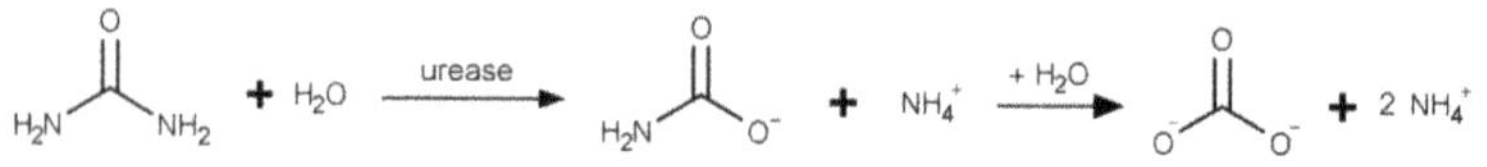

$$H_2N\text{-}C(=O)\text{-}NH_2 + H_2O \xrightarrow{\text{urease}} H_2N\text{-}C(=O)\text{-}O^- + NH_4^+ \xrightarrow{+H_2O} {}^-O\text{-}C(=O)\text{-}O^- + 2\ NH_4^+$$

Scheme 5.3. Urease reaction. Urea is hydrolyzed by urease to carbamate and ammonia. Carbamate is then spontaneously hydrolyzed to carbonate and a second equivalent of ammonia.

Urease catalyzes the hydrolysis of urea to carbamate and ammonia. Carbamate is then spontaneously hydrolyzed to carbonate and a second equivalent of ammonia (Scheme 5.3).[114] Urease is a very fast enzyme, with a reaction rate of up to 20,000 s^{-1} per hexamer.[122] Urease activity can either be measured via the pH change of the solution,[122] by quantification of the produced ammonia with *e.g.* salicylate,[123] or *in situ* via a coupled enzymatic assay with glutamic acid dehydrogenase via its consumption of NADH (see Chapters 9.5 and A.2.5).[124]

5.4 F_oF_1-ATP synthase

F_oF_1-ATP synthase can carry out two important catalytic functions in the form of a reversible chemical reaction. The main biological function of this membrane-bound enzyme is the synthesis of ATP, which is driven by a proton gradient. *Vice versa*, some F_oF_1-ATP synthase can establish a proton gradient by hydrolysis of ATP. F_oF_1-ATP synthases, which can be found in mitochondria, chloroplasts, or bacterial membranes (here, from *Escherichia coli*), consist of two parts: The membrane-bound F_O part and the hydrophilic F_1 part. During catalysis in both parts subunits are rotating against each other. F_oF_1-ATP synthase is therefore considered a motor protein. The unidirectional rotation has been confirmed by observation of attached actin filaments[125] and gold nanorods,[126] and with single molecule Förster resonance energy transfer (FRET).[72]

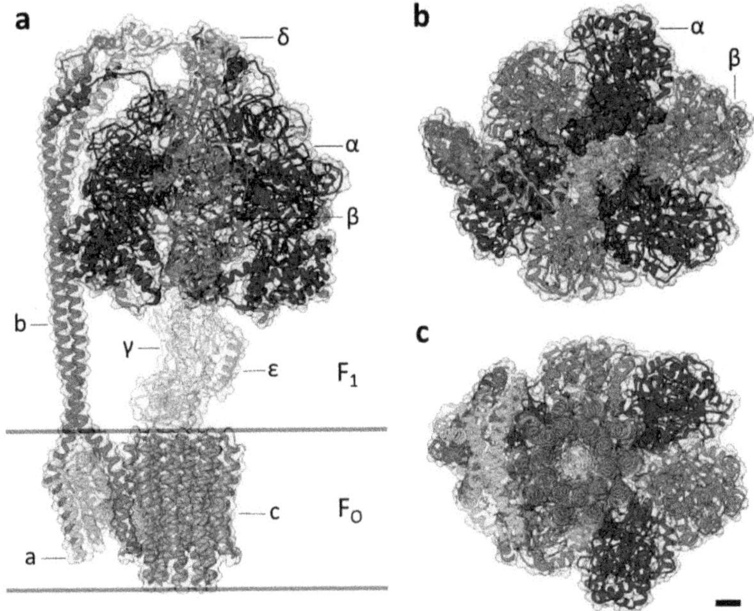

Figure 5.4. Structure of F_oF_1-ATP synthase from *Escherichia coli*. Cryogenic electron microscopy structure from PDB entry 5T4O.[127] Subunits α (blue), β (red), γ (yellow), δ (orange), ε (green), *a* (cyan), *b* (magenta), and *c* (cornflower blue). **a** Side view showing the position of the lipid membrane (grey lines) with embedded F_O part, and the F_1 head part. **b** Top view. **c** Bottom view. Scale bar 2 nm.

The F_O part of ATP synthase, which is involved in proton transport, is made up of the subunits a, b, and c. In *Escherichia coli*, the rotating *c*-ring of the enzyme consists of 10 copies of the *c* subunit and is roughly 5 nm in height (Fig. 5.4). Subunit *a* forms a channel for protons to reach the *c* subunits and, together with the dimeric subunits b_2, connects the two parts F_O and F_1. ATP synthesis or hydrolysis is performed in the F_1 part, which has a subunit composition of $\alpha_3\beta_3\gamma\delta\varepsilon$. Each αβ dimer forms an active site. The subunits γ and ε form the central rotary stalk, which connects the F_1 and F_O parts.[50] The F_1 part is roughly 12 nm in diameter, whereas the whole F_oF_1-ATP synthase is ca. 23 nm in height.[127] Selective labeling of wild-type F_oF_1-ATP synthase is challenging due to its complex structure. Therefore, mutants are often used for labeling and immobilization of F_oF_1-ATP synthase or the F_1 part. Additional cysteine residues,[72,126] His_6 tags,[125,126] and Strep tags[125] have been successfully added by genetic modifications. Fluorescent labels were also directly incorporated without the need of subsequent labeling via genetic modification, *e.g.* as fusions of enhanced green fluorescent protein (EGFP).[72]

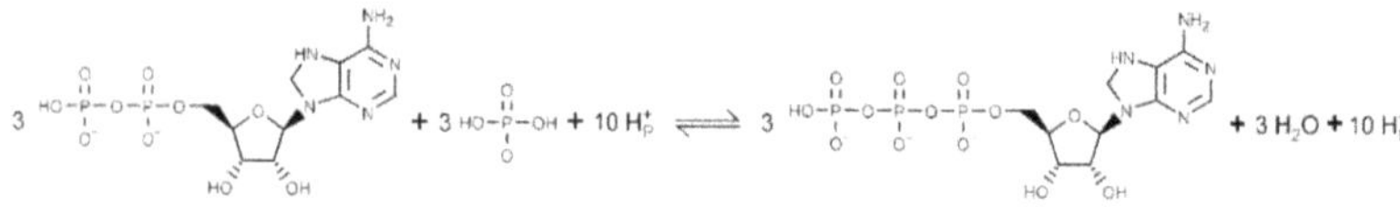

Scheme 5.4. ATP synthesis reaction of F_oF_1-ATP synthase. ADP and phosphate react to ATP while protons are moved from the P- to the N-side of the membrane (main function in living cells). The reverse reaction hydrolyses ATP to generate a proton gradient in the opposite direction to ATP synthesis.

Both parts of F_oF_1-ATP synthase can be considered as individual motors, which can move the rotary domains of each other dependent on the direction of the catalyzed reaction. During catalysis the coupled rotor assembly ($\gamma\varepsilon c_{10}$) rotates against the stator subunits ($\alpha_3\beta_3\delta ab_2$).[50] Within the F_1 part, $\gamma\varepsilon$ is rotating in 120° steps during ATP hydrolysis, which complies with the threefold symmetry of the $\alpha_3\beta_3$ subdomain. During ATP synthesis, protons are moved by the F_O part from the acidic P-side of the membrane to the more alkaline N-side. The F_O part is transporting ten protons per revolution across the membrane since it consists of ten *c* subunits. Therefore, the ratio of synthesized ATP to channeled protons is three to ten (Scheme 5.4).[128] F_oF_1-ATP synthase has a maximum turnover rate of roughly 100 revolutions per second during ATP hydrolysis.[129] It can be inhibited by the substrate analogue β,γ-imidoadenosine-5'-triphosphate (AMPPNP),[72] but is also autoinhibited by its ε subunit. During this regulatory function, the C-terminal domain of the ε subunit is partially inserted into the $\alpha_3\beta_3$ hexamer to halt rotation. ε has a dissociation constant from the soluble F_1 part, which changes over two orders of magnitude in the presence of ATP.[130]

6 Applications of Enzymes

"For close to five thousand years we have made use of microbial enzymes to brew beer and leaven bread. Once the protein catalysts were identified and isolated, many more diverse applications were devised. Today, enzymes are used to diagnose and treat disease, reduce farm waste, enhance textiles and other materials, synthesize industrial and pharmaceutical chemicals, and empower our laundry detergents. But so much more could be achieved if we understood how to build new ones."[131]

Frances Hamilton Arnold

This chapter aims to provide a short summary of the advantages and disadvantages of enzymes in biocatalysis. It focuses on immobilization techniques, which enable interesting collective effects, which will be applied in Chapter 9.

6.1 Biocatalysis

Evolution has been optimizing enzymes for their biological applications over billions of years.[132] This catalytic potential can be applied to organic synthesis, where enzymes can replace (synthetic) chemical catalysts in existing reactions or open up new pathways for novel synthesis routes. If the wild-type enzyme does not have the desired features such as promiscuity (an enzyme's ability to catalyze multiple reactions) or stability in a certain environment, then directed evolution can be applied. In directed evolution, the enzymes are engineered under specific conditions to yield the desired properties in a practical time frame not achievable via natural evolution.[133] As the enzymes themselves, the field of biocatalysis is constantly evolving and inventing new strategies to modify and apply enzymes for technological applications.[134]

Enzymes have several advantages over conventional, chemical catalysts:[135]

- *Activity:* Enzymes have in general high catalytic activity compared to chemical catalysts. The average enzyme has a turnover rate of ~10 s^{-1}, but some enzymes have efficiencies at the diffusion limit.[136] Enzymes are highly effective and can speed up reactions by factors ranging from 10^7 up to 10^{19} times compared to the non-catalyzed reaction.[137]
- *Mild conditions:* Enzymes can be used at relatively mild temperatures, pressures, and in aqueous media, since most enzymes operate at ambient conditions.
- *Waste management:* The disposal of enzymes can also be easier than for some chemical catalysts, since they are biodegradable. The use of aqueous solution is also environmental friendly in comparison to organic solvents.
- *Cascades:* Another major advantage of enzymes is that they can be used in cascades, since they are usually compatible with each other and require similar reaction conditions. This allows complex one-pot syntheses without need for intermediate purification steps.
- *Promiscuity:* Enzymes react with a surprisingly large variety of non-natural substrates, which allows wide applicability of enzymes in synthesis.
- *Selectivity:* Enzymes show high selectivity on a given substrate. Enzymes are enantioselective in addition to their chemo- and regioselectivity. This is caused by the chirality of proteins, since they are naturally made up of L-amino acids.

Despite those phenomenal advantages, enzymes also have disadvantages:[135]

- *Narrow conditions:* Although the conditions for enzymatic catalysis are mild, they are usually narrow and do not allow much change in temperature or pH. Enzymes generally have to be used in aqueous solutions, even though there are many exceptions.

- *Cofactors:* Many enzymes require redox cofactors (*e.g.* NADH, heme, or ATP) which are expensive and usually cannot be replaced.
- *Product inhibition:* Many enzymes decrease their activity with increasing product concentration. This product inhibition is not desirable for biocatalysis, but is necessary *in vivo* where enzymes need to self-regulate their function.
- *Purification:* Enzymes are difficult to purify and to separate from their product due to their small size, which requires ultracentrifugation or ultrafiltration for separation. Immobilization strategies (see next subchapter) can circumvent this problem.
- *Chirality:* Naturally provided enzymes are only available in one enantiomeric form, which only allows the synthesis of one of the possible enantiomeric products.

6.2 Immobilization

As mentioned before, the purification and recovery of enzymes after reactions is in many cases very challenging. Even though, immobilizing enzymes can be challenging and in some cases can reduce enzyme activity, it has the advantage that enzymes can be easily recovered. This allows the use of more expensive enzymes for synthetic applications. Immobilization also enables collective effects, which can only be achieved by densely packed immobilized enzymes, as will be discussed in the following subchapter.

Common templates for enzyme immobilization are inorganic surfaces. If the enzyme is directly bound to the surface, the orientation and number of binding sites is difficult to control, which often leads to reduced activity of the enzyme.[138] Despite these disadvantages, inorganic templates also offer the advantage to control shape, porosity, optical properties and magnetic properties of the overall structure. For example, some metallic templates can be used for plasmonic sensing of the structure's surrounding or position, which can be used to determine protein binding kinetics to the surface[139] or remotely heat the sample.[140] Magnetic materials on the other hand provide the ability for remote control and recovery of the structure. Modern deposition techniques also allow high control over shape and composition of nanomaterials,[141] which can be difficult for organic nanomaterials with the prominent exception of self-assembled DNA nanostructures.[142]

Organic materials are better suited for enzyme immobilization regarding the conservation of activity, since organic materials closer resemble the natural surroundings of enzymes. Examples for organic templates include DNA nanostructures[143] and viruses.[144] In both cases, the template is self-assembled, but DNA nanostructures have the additional advantage as their shape can be preprogrammed, whereas this is only possible in a limited way for viruses via genetic engineering.

A very common type to nanostructure DNA is the use of DNA origamis, which are self-assembled structures consisting of a long scaffold strand (*e.g.* the circular single-stranded DNA of M13 bacteriophages) and short staple strands, which define the shape.[145] Enzymes have been successfully immobilized on DNA origami structures with high precision in the positioning of individual enzymes or even cascades of enzymes.[146–148] This can be achieved by encoding the binding sites into the staple strands. Disadvantages of DNA origamis as enzyme templates are the high price and the need for complex purification steps.[149]

Viruses are also promising templates for enzyme immobilization,[144] since their proteinaceous nature has little effect on enzyme activity and allows various binding strategies. Elongated viruses such as the tobacco mosaic virus[150] and M13 bacteriophages are promising due to their large surface area. Plant viruses and bacteriophages are especially advantageous, since they are safe for human usage and are relatively easily produced and extracted from their natural targets (plants and bacteria). The viruses can also be genetically modified with functionalization sites or even complete enzymes,[151] but the possibilities are limited, since the ability to infect the target should not be altered for a successful modification.

An interesting bacteriophage for the immobilization of enzymes is the M13 bacteriophage (from now on phage). Phages are filamentous bacteria-infecting viruses with their genome encapsulated in form of single-stranded DNA. They have a ca. 1 µm long and 6 nm wide capsid consisting of ~ 2700 copies of the major coat protein pVIII and five copies of the minor coat proteins, which are either specific to the head or tail (Fig. 6.1). This allows genetic modifications on either head, tail or body, which can be used for directed binding of the phage to a surface. Enzyme immobilization is *e.g.* possible via chemical modification of the major coat protein,[152] which allows for dense enzyme packing due to the high number of pVIII proteins per phage. More details on immobilization of enzymes on phages are discussed in Chapter 9.

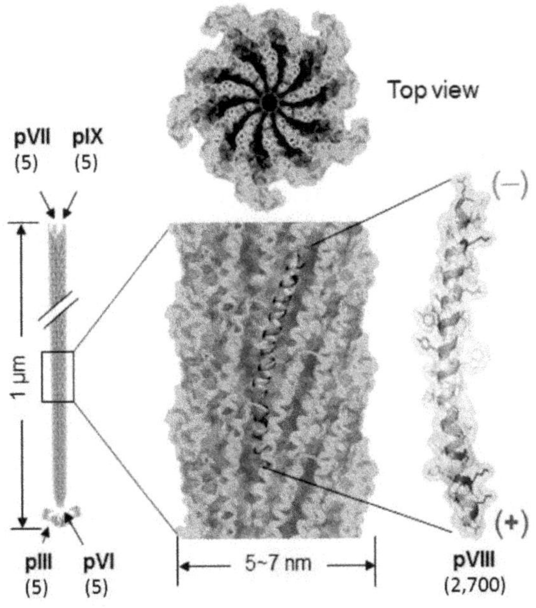

Figure 6.1. Filamentous phage structure. Number of coat proteins for M13 bacteriophage in brackets. Image modified and taken from Ref. 153.

6.3 Collective Effects

Some collective effects involving enzyme immobilization such as the propulsion of self-phoretic swimmers have already been discussed in Chapter 3.1. This chapter will focus on additional effects caused by localizing enzymes. Collective effects of enzymes are well-known, *e.g.* the collective work of myosin in muscles to perform macroscopic motion. It is also a topic of ongoing research, to determine whether free floating enzymes in cells form aggregates called metabolons to increase the efficiency of cascade reactions.[154] Metabolons are an example of substrate channeling. During substrate channeling, the diffusive search of intermediates for the next enzyme in a cascade is shortened to increase the overall efficiency of the cascade. This effect has been proposed for enzyme cascades immobilized on DNA origamis. It should be noted that spatial proximity itself (even at the nanoscale) is not sufficient for substrate channeling, since the intermediates more likely diffuse away from the template than to the next enzyme in the cascade. The effect of enhanced activity of an enzyme cascade-reaction on DNA origamis has been explained with a local pH changes due to the high charge of the DNA, rather than the initially suggested spatial proximity effects.[155–157]

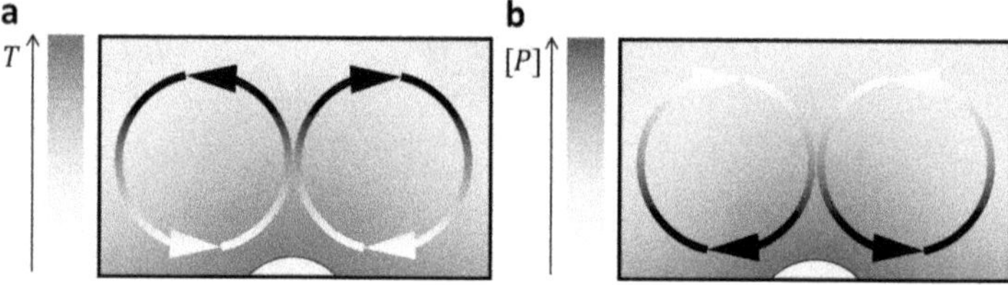

Figure 6.2. Possible driving forces of (enzymatic) micropumps. A patch of catalyst (*e.g.* enzyme) (yellow) is immobilized on the bottom of the container. Density differences lead to convective flow (arrows). **a** Thermal buoyancy. **b** Solutal buoyancy.

Enzymes can also be used to collectively drive fluids and create micropumps.[158–160] When many enzymes are immobilized inhomogeneously at the wall of a container, they can generate a flow in the container. The density differences, which drive convective flows, can be caused by two different mechanisms (thermal and solutal buoyancy) depending on the catalyzed reaction and the reaction rate (Fig. 6.2).[159] Thermal buoyancy dominates, if the catalyzed reaction is exothermic and a temperature gradient can build up, which drives the convection. Due to the localized reaction, concentration gradients of substrates and products can also build up within the container. These concentration gradients correspond to density gradients and lead to solutal buoyancy, if the substrate and product have different densities. Both mechanisms convert chemical energy into motion. The devices are called micropumps, since the containers are usually on the millimeter to micrometer scale. Micropumps have potential applications in autonomous microfluidic devices. The immobilization of enzymes allows the use of biological fuels for the micropump, so that a biomedical sample itself could provide the fuel to pump itself, for instance through a lab on a chip device. A novel fabrication strategy for enzymatic micropumps, which yielded the fastest flows to date, is discussed in Chapter 9.

7 Misinterpretation of Enzyme FCS

The publication presented in this chapter, "Diffusion Measurements of Swimming Enzymes with Fluorescence Correlation Spectroscopy", was published in August 2018 in *Accounts of Chemical Research* as part of the special issue "Fundamental Aspects of Self-Powered Nano- and Micromotors". Prior to this publication, there were several experimental reports supporting the active enzyme diffusion hypothesis (see Tab. 7.1) and few theoretical reports, which were critical of it.[28,55] The following publication presents four artefacts, which can lead to misinterpretation of enzyme FCS data and could therefore falsify the active enzyme diffusion hypothesis:

- Oligomer and Subunit Dissociation
- Surface Binding
- Conformational Change
- Fluorophore Quenching

Theoretical estimations and experimental examples for all artefacts were presented. Prior to this publication, alkaline phosphatase was believed to be the enzyme with the highest diffusion enhancement,[8] which was falsified with FCCS measurements presented in this publication, which revealed fluorophore quenching to be the cause for these earlier misinterpretations. For the enzymes F_1-ATPase and alkaline phosphatase active diffusion could be ruled out within the accuracy of the performed FCCS measurements.

Contributions: J.-P. G. wrote the first draft of the manuscript. All authors contributed to the finished manuscript. J.-P. G. labeled, purified and characterized all samples of alkaline phosphatase including activity measurements. M. B. prepared all samples of ATP synthase. J.-P. G. assisted M. B. with the FCCS measurements. J.-P. G. and M. B. analyzed the data. J.-P. G. performed all simulations.

Acc. Chem. Res. **51**, 1911–1920 (2018)

Diffusion Measurements of Swimming Enzymes with Fluorescence Correlation Spectroscopy

Jan-Philipp Günther,[†,‡] Michael Börsch[§,] and Peer Fischer[†,‡]*

[†]Max Planck Institute for Intelligent Systems, 70569 Stuttgart, Germany

[‡]Institute of Physical Chemistry, University of Stuttgart, 70569 Stuttgart, Germany

[§]Jena University Hospital, Friedrich-Schiller University Jena, 07743 Jena, Germany

**E-mail: michael.boersch@uni-jena.de*

7.1 Conspectus

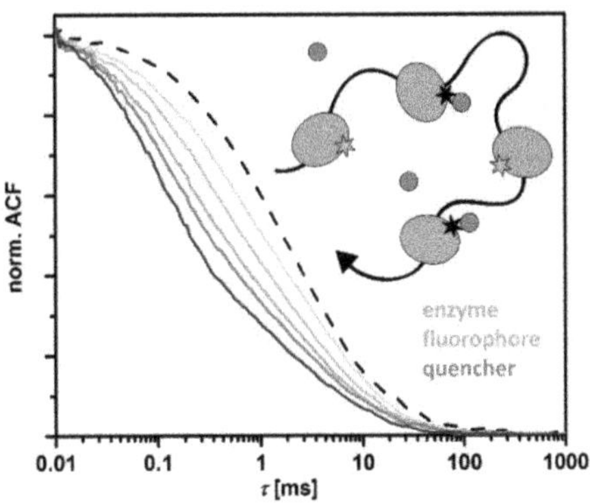

Self-propelled chemical motors are chemically powered micro- or nano-sized swimmers. The energy required for these motors' active motion derives from catalytic chemical reactions and the transformation of a fuel dissolved in the solution. While self-propulsion is now well established for larger particles, it is still unclear, if enzymes – nature's nanometer-sized catalysts – are potentially also self-powered nanomotors. Because of its small size, any increase in an enzyme's diffusion due to active self-propulsion must be observed on top of the enzyme's passive Brownian motion, which dominates at this scale. Fluorescence correlation spectroscopy (FCS) is a sensitive method to quantify the diffusion properties of single fluorescently-labeled molecules in solution. FCS experiments have shown a general increase in the diffusion constant of a number of enzymes, when the enzyme is catalytically active. Diffusion enhancements after addition of the enzyme's substrate (and sometimes its inhibitor) of up to 80 % have been reported, which is at least one order of magnitude higher than what theory would predict. However, many factors contribute to the FCS signal and in particular the shape of the autocorrelation function, which underlies diffusion measurements by fluorescence correlation spectroscopy. These effects need to be considered to establish if and by how much the catalytic activity changes an enzyme's diffusion.

We carefully review phenomena that can play a role in FCS experiments and the determination of enzyme diffusion, including the dissociation of enzyme oligomers upon interaction with the substrate, surface binding of the enzyme to glass during the experiment, conformational changes upon binding, and quenching of the fluorophore. We show that these effects can cause changes in the FCS signal that behave similar to an increase in diffusion. However, in the case of the enzymes F_1-ATPase and alkaline phosphatase, we demonstrate that there is no measurable increase in enzyme diffusion. Rather, dissociation and conformational changes account for the changes in the FCS signal in the former and fluorophore quenching in the latter. Within the experimental accuracy of our FCS measurements, we do not observe any change in diffusion due to activity for the enzymes we have investigated.

We suggest useful control experiments and additional tests for future FCS experiments that should help establish if the observed diffusion enhancement is real or if it is due to an experimental or data analysis artefact. We show that fluorescence lifetime and mean intensity measurements are essential in order to identify the nature of the observed changes in the autocorrelation function. While it is clear from theory that chemically active enzymes should also act as self-propelled nanomotors, our FCS measurements show that the associated increase in diffusion is much smaller than previously reported. Further experiments are needed to quantify the contribution of the enzymes' catalytic activity to their self-propulsion. We hope that our findings help to establish a useful protocol for future FCS studies in this field and help establish by how much the diffusion of an enzyme is enhanced through catalytic activity.

7.2 Introduction

The passive Brownian diffusion of molecules and colloids is a well-known phenomenon and its measurement provides valuable information on their shape, hydrodynamic radius, or the rheology of their environment. That colloids and even enzymes may also actively swim in addition to their passive Brownian motion is a much newer concept[7] and one that has potentially profound implications for active matter and the dynamics and interactions of enzymes. Self-propulsion in particles is well-established and has been observed for many chemically-active colloids of different shapes and covered with various catalysts.[37,161,162] One interesting question is whether catalytic reactions that propel colloids can also cause self-propulsion in very small nanoparticles[35,38,163] or enzymes.

Kapral and colleagues formulated a theory and find that catalytically active Ångström-sized nanoparticles exhibit enhanced diffusion and consequently swim.[31] Similar theoretical results have been proposed for catalytically active enzymes, which typically have a hydrodynamic radius of a few nanometers. While measurements of enzyme diffusion,[11,7,23,8–10,12] which are all based on fluorescence correlation spectroscopy (FCS) measurements, report increases in the diffusion constant of enzymes upon catalytic activity, there is as yet no agreement between experiment and theory. To the best of our knowledge there is no single unified theoretical explanation on how enzymes could actively swim.[55] While enzymes could in principle show enhanced diffusion and swimming, Bai and Wolynes conclude that the effect is smaller than recent experiments suggest.[28] Here, we carefully examine effects that contribute to changes in FCS signals and how they can affect the determination of the diffusion constant. In the case of the catalytic F_1 domain of F_oF_1-ATP synthase we can clearly show, that FCS measurements – which on first inspection suggest increased diffusion – are in fact due to a number of effects that are not related to self-propulsion. These observations suggest that a number of additional controls are necessary in model-based analysis of FCS measurements of (enhanced) enzyme diffusion, which we describe here.

7.3 Experiment and Theory of the Enhanced Diffusion of Active Enzymes

Fluorescence correlation spectroscopy (FCS), which we discuss below, has been used to study the diffusive behavior of enzymes. In Tab. 7.1 we have listed in chronological order measurements that report a change in the diffusion constant of an enzyme as a function of its activity. When the enzyme is immersed in a solution containing its substrate, an increase in the translational diffusion from 10 % to 80 % has been reported. Remarkably, a number of experiments suggest that even the interaction of an enzyme with a competitive inhibitor can cause increases in the diffusion constant. The highest enhancement has been reported for the enzyme alkaline phosphatase in solutions containing its substrate p-nitrophenyl phosphate.[8] To the best of our knowledge eight enzymes have been studied thus far and only triose phosphatase isomerase has not shown any increase in its diffusion constant (see Tab. 7.1).

Table 7.1. Reports of enhanced enzyme diffusion determined with FCS measurements in chronological order. D/D_0 denotes the maximum ratio of translational diffusion constant with (D) and without (D_0) interacting substrate/inhibitor reported. Data marked by * was measured with STED-FCS and data marked with ‡ was extracted from the corresponding figures.

enzyme (*species* or source)	substrate or [competitive inhibitor]	D/D_0	Ref.
F_1-ATPase (*Escherichia coli*)	adenosine triphosphate	1.23	11
F_1-ATPase (*Escherichia coli*)	[imidoadenosine-5'-triphosphate]	1.14	11
urease (*Canavalia ensiformis*)	urea	1.28	7
catalase (*Bos taurus*)	hydrogen peroxide	1.45	158
urease (*Canavalia ensiformis*)	urea	1.3‡	8
catalase (*Bos taurus*)	hydrogen peroxide	1.3‡	8
alkaline phosphatase (bovine)	p-nitrophenyl phosphate	1.8‡	8
triose phosphate isomerase (*S. cerevisiae*)	D-glyceraldehyde 3-phosphate	1.0‡	8
fructose bisphosphate aldolase (rabbit)	fructose-1,6-bisphosphate	1.3‡	9
fructose bisphosphate aldolase (rabbit)	[pyrophosphate]	1.2‡	9
urease (*Canavalia ensiformis*)	urea	1.5*‡	10
acetylcholinesterase (*E. electricus*)	acetylcholine	1.2*‡	10
hexokinase (*Saccharomyces cerevisiae*)	D-glucose	1.38	12

The theoretical explanations for these observations differ widely. Riedel *et al.* suggest that a "chemoacoustic effect", during which the chemical energy of the substrate turnover is converted to pressure waves, is propelling the enzyme.[8] This explanation has been questioned in a number of publications.[28,55–57] Golestanian calculated that temperature increase due to exothermicity of the reaction would also cause a comparable increase of the enzyme diffusion.[55] More puzzling is that even endothermic enzyme reactions (for aldolase) were found to cause increased diffusion. Illien *et al.* concluded this could be caused by fluctuation-induced hydrodynamic coupling of the enzyme's conformational changes with the surrounding solution.[9] Bai and Wolynes on the other hand suggested that photophysics is the cause for the observed change in the transit time from which one may erroneously deduce an apparent change of the diffusion constant.[28] In what follows, we will briefly review how diffusion constants are obtained from FCS measurements and then examine what this implies for measurements of enhanced diffusion in active enzymes.

7.4 Fluorescence Correlation Spectroscopy

In fluorescence correlation spectroscopy (FCS) the diffusion of single fluorophores through a diffraction-limited confocal volume (Fig. 7.1A) is observed and analyzed to yield translational diffusion coefficients as well as photophysical parameters.[164] A microscope setup for FCS includes an excitation laser, which is reflected by a dichroic mirror and focused through an objective into the sample solution. The confocal detection volume is usually in the range of one fL depending on the size of the pinhole and the laser beam diameter. The fluorophore concentration is selected to yield on average one to five fluorophores in the confocal volume (corresponding to nM or pM concentrations). The emitted fluorescence of the single fluorophores traversing the confocal volume is then collected and recorded as an intensity time trace with avalanche photodiodes (APD) for time-correlated single photon counting (TCSPC) (Fig. 7.1B). The analysis of the photon time trace involves calculating the autocorrelation function (ACF) $G(\tau)$ of the APD signal $I(t)$.

$$G(\tau) = \frac{\langle I(t) \cdot I(t+\tau) \rangle}{\langle I(t) \rangle^2} \tag{7.1}$$

The resulting autocorrelation curve can then be fitted with the following function (Fig. 7.1C):

$$G(\tau) = G(\tau)_D \, G(\tau)_T \tag{7.2}$$

where $G(\tau)_D$ represents the contributions of fluorophore diffusion to the ACF and $G(\tau)_T$ the contribution of triplet states. Most FCS analyses assume only translational diffusion of the fluorophore, which means the autocorrelation function takes the form $G(\tau) \approx G(\tau)_D$:[73]

$$G(\tau)_D = \frac{1}{N}\left(\frac{1}{1+\tau/\tau_D}\right)\left(\frac{1}{1+(\omega_0/z_0)^2\,\tau/\tau_D}\right)^{\frac{1}{2}}, \tag{7.3}$$

where N is the average number of fluorophores in the confocal volume, τ the correlation time, τ_D the translational diffusion time, ω_0 the radius in the radial direction and z_0 the radius in axial direction. However, all fluorophores exhibit "blinking" and therefore $G(\tau)_T$, which describes the triplet state (one of the possible "dark states") of the fluorophore, has to be considered, too.[165]

$$G(\tau)_T = 1 + \frac{T}{(1-T)}\exp\left(-\frac{\tau}{\tau_T}\right) \tag{7.4}$$

Here T is the fraction of molecules in the triplet state and τ_T the triplet relaxation time. In the absence of other effects, the translational diffusion coefficient D of the fluorescently labeled species can be calculated with τ_D obtained from the fit of $G(\tau)$ Eq. (7.2), since:

$$D = \frac{{\omega_0}^2}{4\,\tau_D} \tag{7.5}$$

Hence, FCS analysis requires calibration of the confocal volume parameters ω_0 and z_0 for each measurement. This means that freely diffusing dyes are measured in water, for which the diffusion constant has been determined with independent high accuracy methods, such as pulsed field gradient NMR.[82] Therefore, diffusion constants in FCS are determined relative to reference measurements. Since classical FCS only determines transition times through the confocal volume via τ_D, it cannot easily distinguish between random and directed motion. For this more elaborate methods such as dual-focus FCS[166] and STED-FCS[10] may be needed.

Any effect that changes the fluorescence lifetime of the fluorophore or the labeled species must be considered to ensure that only τ_D enters the determination of the diffusion constant. However, even seemingly unimportant effects that occur at much longer timescales, such as adhesion of the enzymes to glass substrates (which we will discuss below for F_1-ATPase), can affect the form of the autocorrelation curve and hence the deduced diffusion constant.

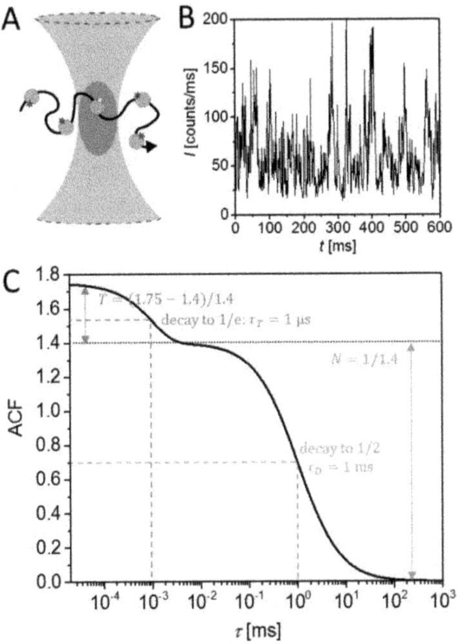

Figure 7.1. FCS principle. A) Scheme of the confocal volume of an FCS setup with a fluorophore trajectory. B) Intensity time trace with individual photon bursts seen in the APD signal. C) Calculated autocorrelation function (ACF) (Eq. 7.2) with the diffusion term at longer correlation times (green) and the contributions of the triplet state at shorter correlation times (red).

7.5 Enhanced Diffusion of F_1-ATPase

An increase of an observed diffusion constant of an enzyme has already been published by one of us twenty years ago for the soluble F_1 domain of F_oF_1-ATP synthase exposed to its substrate adenosine triphosphate (MgATP) as well as an inhibitor.[11] In the case of the inhibitor, it has been possible to obtain additional electron microscope structural data of the enzyme-inhibitor complex that shows that the enzyme undergoes a conformational change upon binding of the inhibitor and reduces its size by about 15 %.[167] A smaller hydrodynamic radius necessarily results in an increase in diffusion. So, here the increase in diffusion upon binding is due to a conformational change. However, further effects also play a role, as new measurements of F_1-ATPase labeled with Alexa Fluor 488 with MgATP as substrate in Fig. 7.2 show. The autocorrelation curves of the F_1-ATPase are seen to show faster decays in Fig. 7.2B over the course of the experiment. The explanation, however, is not a faster diffusion of the enzyme. Rather, the low concentration of the labeled enzyme, which is typical for FCS measurements, is close to the dissociation constants of the subunits of the multi-domain enzyme.[130] Fig. 7.2B thus might imply a 3.3-fold increase of the diffusion constant, but this actually arises, because of the dissociation of the ε subunit where the fluorophore is bound from the rest of the F_1-ATPase. Since the subunit with the fluorophore is significantly smaller than the parent enzyme, it diffuses much faster. Calculations with the software Hydropro[95] based on the Protein Data Bank crystal structures 5T4O[127] for F_1-ATPase and 1BSN[168] for the ε subunit suggest a 2.9-fold diffusion constant increase and support this interpretation of the FCS data.

In addition to the two aforementioned effects, one also finds that F_1-ATPase binds to the glass substrate of the observation chamber, as is schematically depicted in Fig 7.2C. This also results in a decrease of the fluorescence intensity over time due to the diminishing number of fluorescent molecules in the confocal detection volume, as is seen in the experimental data of Fig. 7.2D. While the kinetics of adhesion is slow, it can nevertheless affect the shape of the autocorrelation function. The reason is that there are always some free unattached dye molecules in solution. Due to the adhesion and hence loss of enzymes from solution, the ratio of labeled enzymes to free dye molecules changes over time, which shifts the ACF to shorter correlation times. Again, there is no real speed-up of the enzyme. In general, a number of effects simultaneously play a role and affect the autocorrelation function and these must all be considered when deducing enzyme diffusion properties. We now elaborate the impact of these three effects, shown for F_1-ATPase, as well as the impact of fluorescence quenching on FCS diffusion measurements of enzymes in more detail.

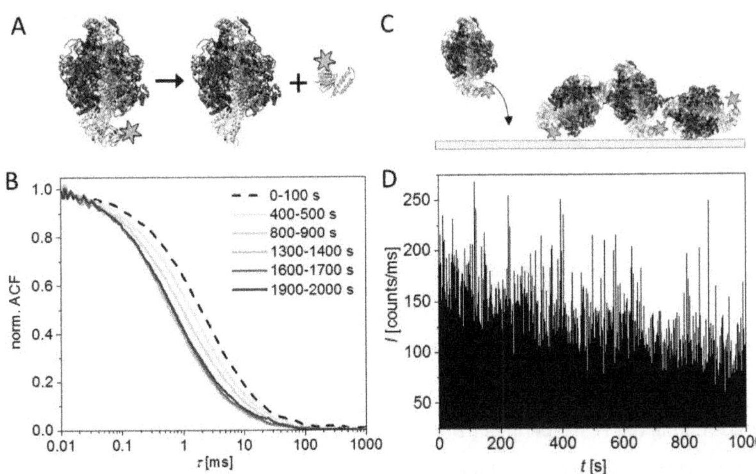

Figure 7.2. Artefacts observable by FCS of catalytically active F_1-ATPase. A) Sketch of the dissociation of the labeled ε subunit (green) from the remaining F_1-ATPase with subunits α (blue), β (red), γ (yellow) and δ (orange). Fluorophore positions are marked by green asterisks. B) Time lapse of FCS curves due to dissociation of the Alexa488-labeled ε subunit in buffer (20 mM tricine-NaOH, 20 mM succinate, 0.6 mM KCl, 2.5 mM $MgCl_2$, pH 8.0)[72] with 1 mM MgATP. C) Sketch of surface binding of F_1-ATPase. D) Fluorescence intensity time trace of labeled F_1-ATPase, which decreases due to surface binding. Structure of the F_1-ATPase and ε subunit were obtained from Protein Data Bank entries 5T4O[127] and 1BSN[168] and depicted with Chimera.[99]

7.6 Effects and Artefacts in FCS Measurements of Protein Diffusion

Oligomer and subunit dissociation. Apart from F_1-ATPase, many enzymes, including for instance aldolase, comprise different non-covalently bound subunits to form active hetero- or homo-multimers. Since FCS is performed at very low concentrations in the nM to pM range, subunits and oligomeric forms of the enzymes can dissociate, if the concentration gets too close to the dissociation constants (Fig. 7.3A-B). This leads to a mismatch between the apparent diffusion constant and the expected hydrodynamic radius. The even more severe problem for active enzyme diffusion measurements is a substrate-induced dissociation of enzymes. This has been observed for aldolase[109] and for F_1-ATPase.[130] Therefore diffusion increase in such systems upon substrate addition might be dominated by size change of the enzyme caused by dissociation.

Surface binding. Binding of proteins to the glass cover slides used in FCS experiments impacts the effective diffusion time. When a protein is fluorescently labeled, some unreacted dye molecules remain in solution. Even elaborate purification methods can only reduce the amount of free dye, but cannot remove it completely. If the labeled protein binds to the glass surface during the FCS experiment the ratio of free to bound dye molecules changes. This effect is represented in the ACF through the sum of multiple diffusion terms with different N_i and $\tau_{D,i}$ for each fluorescent species i in solution:[73]

$$G(\tau) = \frac{1}{\sum_i N_i^2} \sum_i \left(N_i \left(\frac{1}{1 + \tau/\tau_{D,i}} \right) \left(\frac{1}{1 + (\omega_0/z_0)^2 \, \tau/\tau_{D,i}} \right)^{\frac{1}{2}} \right) \tag{7.6}$$

Since surface binding of the slower diffusive species (*i.e.* labeled proteins) leads to a higher relative amount of the remaining faster diffusive free dyes, an apparent diffusion coefficient increase can be observed over time (Fig. 7.3C-E). A decrease of the fluorescence intensity or fluorescence anisotropy over time may signal that the number of labeled proteins in solution is reduced.

Conformational change. Based on the induced fit model, enzymes can adapt their conformation upon substrate binding, which in turn leads to a hydrodynamic radius change. Hexokinase for instance is a well-studied example that shows conformational changes. Upon binding of its substrate glucose, the domains of the protein close by about 17°, which includes movements of over 1 nm of the polypeptide backbone.[47] Using the software Hydropro[95] the diffusion enhancement of such conformational changes can be estimated from crystal structures.[47,48] In the case of hexokinase this leads to a radius of gyration decrease of 3.6 %. This effect has also been observed for F_1-ATPase upon binding of the competitive inhibitor imidoadenosine-5'-triphosphate (MgAMPPNP), where the size decrease is 15 %[167] (compare Tab. 7.1, Fig. 7.3F-G). Later it has been revealed that dissociation of the δ and ε subunits occurs during FCS measurements,[130,169] which contributes to the apparent diffusion enhancement (Fig. 7.2A-B).

Fluorophore quenching. Quenching of the fluorophore used for labeling can complicate the analysis of FCS measurements and may erroneously lead one to conclude that there is a diffusion enhancement, if the triplet relaxation time τ_T (or the correlation times of other photophysical dark states) are comparable to the translational diffusion time τ_D. When a quencher interacts with the exited states of the fluorophore (*i.e.* singlet S_1 or triplet T_1, Fig. 7.3I) through quenching, the rate constants k_X between the different states change depending on the quencher concentration. Accordingly, T and τ_T also change with the quencher concentration.[165]

$$k_X = k_{X,0} + k_{X,Q}[Q] \quad (X = 10, ISC, T) \tag{7.7}$$

$$T = \frac{k_{01}k_{ISC}}{k_{01}(k_T + k_{ISC}) + k_{10}k_T} \tag{7.8}$$

$$\tau_T = \frac{k_{01} + k_{10}}{k_{01}(k_T + k_{ISC}) + k_{10}k_T} \tag{7.9}$$

If the substrate or product of the analyzed enzymatic reaction is a quencher, this can lead to an ostensibly activity-dependent diffusion increase. Typical quenchers are for instance amines, halogen ions, hydrogen peroxide, nitro compounds and oxygen.[73] These quenchers

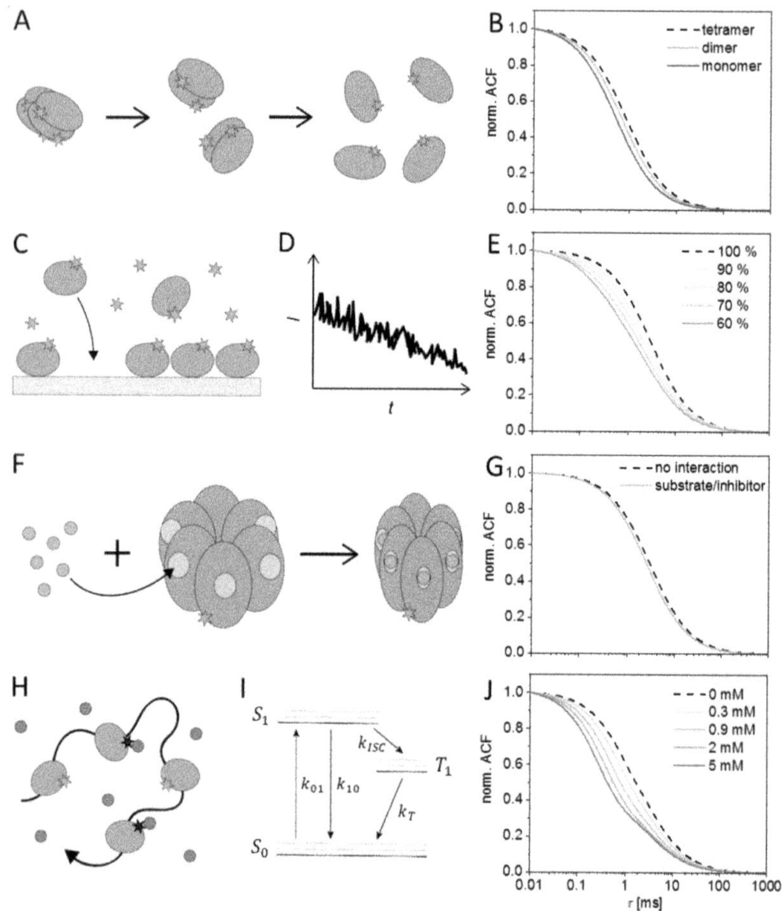

Figure 7.3. Sketches of effects and corresponding simulated autocorrelation functions ($\tau_{D,0} = 3$ ms, $\omega_0 = 0.4$ µm, $z_0 = 4$ µm and $N = 0.5$, if not stated otherwise). A-B) Multimer dissociation of aldolase based on diffusion constants calculated with Hydropro[95] from Protein Data Bank Entry 1ZAH.[113] C-E) Free dyes and surface binding ($\tau_{D,1} = 3$ ms, $\tau_{D,2} = 0.3$ ms). F-G) Conformational change of F_1-ATPase with 15 % radius decrease.[167] H-J) transient quenching and Jablonski diagram with parameters $k_{01} = 10^8\ \mathrm{s}^{-1}$, $k_{10} = 2 \cdot 10^8\ \mathrm{s}^{-1} + 10^{11}\ \mathrm{M}^{-1}\mathrm{s}^{-1}[Q]$, $k_{ISC} = 10^3\ \mathrm{s}^{-1} + 3 \cdot 10^6\ \mathrm{M}^{-1}\mathrm{s}^{-1} \cdot [Q]$ and $k_T = 10^3\ \mathrm{s}^{-1} + 10^5\ \mathrm{M}^{-1}\mathrm{s}^{-1}[Q]$.

can be found in some buffer solutions and sometimes even as substrates for enzymatic reactions, and have been used in previous studies of enhanced enzyme diffusion. The fluorescence lifetime τ_F of the labeled protein is an indicator for possible quenching processes during FCS experiments, since it is also dependent on the rate constants.

$$\tau_F = \frac{1}{k_{10} + k_{ISC}} \tag{7.10}$$

If a pulsed laser is used in the FCS setup, τ_F of the fluorophore can also be extracted from the TCSPC data (via the so-called microtime histogram) and be fitted with an exponential decay function.

$$I(t) = I(0)\exp\left(-\frac{t}{\tau_F}\right) \tag{7.11}$$

Here I is the photon count, t the arrival time of the photon with respect to the laser pulse and τ_F the fluorescence lifetime.

7.7 Suggestions for Future FCS Experiments of Enzyme Diffusion

In order to identify the reasons for the observed changes of the FCS autocorrelation functions, we would like to suggest the following control experiments. It is highly advisable to analyze the intensity time trace for each experimental condition tested with FCS. Many artefacts like aggregation, dissociation and surface binding can be identified through inspection of the intensity time trace recorded for at least 20 min.

For most enzyme systems it is necessary that the sample is purified after fluorescence labeling, for instance by FPLC (fast protein liquid chromatography). Using only short single-use spin-columns for size exclusion chromatography may not suffice. Proper purification reduces the amount of free dye molecules during the FCS experiment and thus improves the fit of the autocorrelation function. Ideally, more advanced FCS techniques should be used that can distinguish between bound and free fluorophore molecules. For instance polarization-resolved simultaneous anisotropy recording multidetector-FCS[170] allows one to check for unrestricted fluorophore mobility.

In all FCS measurements it is essential that a photostable fluorophore with a high fluorescence quantum yield > 0.8, good solubility, high absorbance with $\varepsilon > 70{,}000\ M^{-1}cm^{-1}$, and reasonably long fluorescence lifetime in the range of 3 to 4 ns is used to achieve a high signal-to-noise ratio. It is also advisable to perform fluorescence lifetime measurements of the fluorophore during the FCS experiment or under similar conditions to check for quenching effects. This we are now going to demonstrate with advanced FCS measurements of alkaline phosphatase, which clearly reveal the artefacts that lead one to erroneously conclude that alkaline phosphatase shows a dramatic enhancement in its diffusion when it is turning over substrate.

7.8 Fluorescence Quenching of Alkaline Phosphatase

Alkaline phosphatase is an enzyme, which catalyzes dephosphorylation at elevated pH. Alkaline phosphatase from bovine intestinal mucosa has shown the highest reported diffusion increase of active enzymes so far (excluding the results of F_1-ATPase by subunit dissociation in Fig. 7.2B). The substrate of alkaline phosphatase used in the study by Riedel *et al.*[8] is p-nitrophenyl phosphate (NPP), which is converted by phosphatase to p-nitrophenol and phosphate. We revisited this enzyme, but focused on fluorescence lifetime measurements, which have not been reported in the initial study. First, we prepared a highly purified sample of alkaline phosphatase labeled with a photostable, bright and highly soluble dye, which we then analyzed with a custom-built FCS setup. A key feature of our setup is that it allows us to simultaneously record the FCS and fluorescence lifetime of the fluorophore bound to the protein. We first describe the setup and the labeling procedures, before we discuss the FCS measurements of phosphatase with NPP and an additional substrate fructose-1,6-bisphosphate (FBP).

FCS setup used in this study. A custom-designed modified confocal microscope was used to record fluorescence as previously described.[170,171] Briefly, a 40 MHz ps-pulsed laser (PicoTA 490; Picoquant) was used for linearly-polarized excitation at 488 nm with 100 µW. A dichroic beam splitter (H 488 LPXR; all filters from AHF) directed the laser through a water immersion objective (UPlanSApo 60XW with N.A. 1.2; Olympus) into the sample positioned in an inverted microscope on a glass cover slide. The fluorescence light was then guided through a 150 µm pinhole and was spectrally split by a dichroic mirror (T 585 LPXR) into two spectral ranges (500 to 570 nm with ET535/70M bandpass filter and λ > 605 nm with 594 LP edge basic longpass filter). Each spectral range was then divided into its parallel and its perpendicular polarization using polarizing beam splitter cubes. Four single photon-counting avalanche photodiodes (2x SPCM-AQRH-14-TR and 2x SPCM-AQRH-14; Excelitas Technologies Corp., Canada) recorded the arrival times of photons using synchronized TCSPC electronics (SPC-154; Becker & Hickl). The analysis of the fluorescence intensity, fluorescence lifetime and the calculation of the autocorrelation functions were performed with software from Becker & Hickl.

Fluorescence labeling of alkaline phosphatase. Alkaline phosphatase from bovine intestinal mucosa (Sigma-Aldrich) was fluorescently labeled with an excess of Alexa Fluor® 488 succinimidyl ester (Life Technologies) in carbonate buffer (0.1 M, pH 8.9) for 1 h at room temperature. Purification of the labeled protein was performed by conventional methods using Microcon® 10 (Merck Millipore) molecular weight cutoff filter and Micro Bio-Spin™ 6 (Bio-Rad) mini size exclusion column. Additionally FPLC purification was performed using a HiPrep™ 16/60 Sephacryl® S-300 HR (GE Healthcare) column in phosphate-buffered saline prepared from Gibco™ PBS Tablets (Thermo Fisher Scientific) followed by Amicon® Ultra-4 10k (Merck Millipore) molecular weight cutoff filtration and Micro Bio-Spin™ 6 (Bio-Rad) for buffer change to 100 mM Tris buffer (pH 9.0, 1 mM $MgCl_2$, 20 µM $ZnCl_2$). All subsequent experiments were performed in this Tris buffer. The labeling efficiency was measured using UV/VIS spectroscopy (Cary 4000, Varian) yielding 27 % labeling per phosphatase dimer (extinction and correction coefficients from Sigma-Aldrich and Life Technologies product information sheets). The catalytic activity of labeled phosphatase was determined spectroscopically using a phosphate assay[104] with malachite green (Sigma-Aldrich) for FBP (Sigma-Aldrich) and direct color change observation[8] for NPP (Sigma-Aldrich). The results of the Michaelis-Menten fits were $k_{cat} = 1360 \pm 330\ \mathrm{s}^{-1}$ and $K_M = 30.3 \pm 16.3\ \mu\mathrm{M}$ per produced phosphate from FBP and $k_{cat} = 3170 \pm 60\ \mathrm{s}^{-1}$ and $K_M = 58.3 \pm 2.1\ \mu\mathrm{M}$ per produced p-nitrophenol from NPP.[172,173]

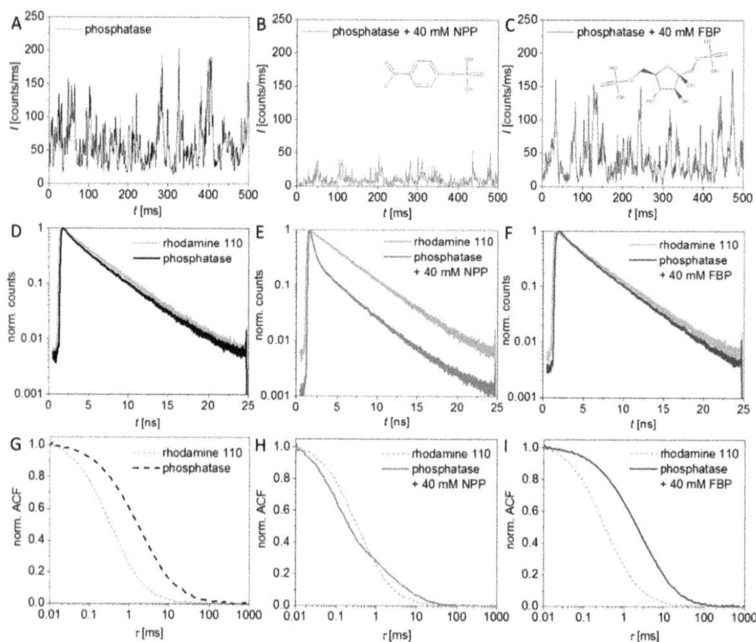

Figure 7.4. Substrate influence on phosphatase fluorescence. Controls without substrate (A, D, G), with NPP (B, E, H) and with FBP (C, F, I) with rhodamine 110 as reference. A-C) Segment of the intensity time traces with insets of the substrate structures. D-F) Fluorescence lifetime measurements extracted from the same measurements as in A-C). G-I) Autocorrelation functions calculated from 100 s recordings taken from the 20 min fluorescence intensity time traces.

Fluorescence lifetime and quenching results for alkaline phosphatase. Fluorescence experiments were performed with 60 pM labeled phosphatase and 2 to 40 mM substrate (FBP and NPP, respectively). Fig. 7.4A-C present exemplary sections of the fluorescence intensity time traces without and with 40 mM substrate. While the fluorescence intensity did not change after addition of FBP, it was strongly reduced, when NPP was present. The fluorescence lifetime, which was calculated from the same photon time traces since a pulsed laser has been used, was also altered to a large degree after the addition of NPP (Fig. 7.4E). The normalized autocorrelation function of phosphatase with NPP was shifted to shorter correlation times, which at first glance suggests an apparent diffusion increase to a diffusion time even shorter than the one of the reference fluorophore rhodamine 110 (Fig. 7.4H). Without any further control experiments this would (incorrectly) suggest a dramatic increase of the diffusion of phosphatase upon substrate conversion of NPP. However, the changes of the fluorescence lifetime and the loss of fluorescence intensity indicate that there is fluorescence quenching of the dye attached to the protein and indeed, nitro compounds are known to be fluorescence quenchers.[174] To further elaborate whether fluorescence quenching was the cause for the apparent diffusion increase, we also measured phosphatase in the presence of the non-quenching substrate fructose-1,6-bisphosphate (FBP). Nearly no differences in diffusion time were observed when FBP was used instead of NPP (Fig. 7.4G and I).

To quantify the fluorescence quenching by NPP we performed different fluorescence experiments with a concentration series for both substrates (Fig. 7.5). For FBP we observed no increase of the diffusion constant, but a decrease (Fig. 7.5A). Fluorescence lifetimes and molecular brightness (*i.e.* average fluorescence intensity per dye molecule, calculated as $\bar{I}/N$) from the concentration series of FBP were constant and well-behaved (Fig. 7.5C-D). The Stern-Volmer plot exhibited no dependency of $\bar{I}_0/\bar{I}$ and $\bar{\tau}_{F,0}/\bar{\tau}_F$ on the concentration of FBP indicating that no fluorophore quenching was present (Fig. 7.5E). The results for NPP however differed dramatically from the ones of FBP, even though both substrates for phosphatase have comparable reaction rates ($k_{cat} = 1360\ \mathrm{s}^{-1}$ for FBP and $3170\ \mathrm{s}^{-1}$ for NPP). The autocorrelation functions of phosphatase with NPP exhibited a strong dependency on the

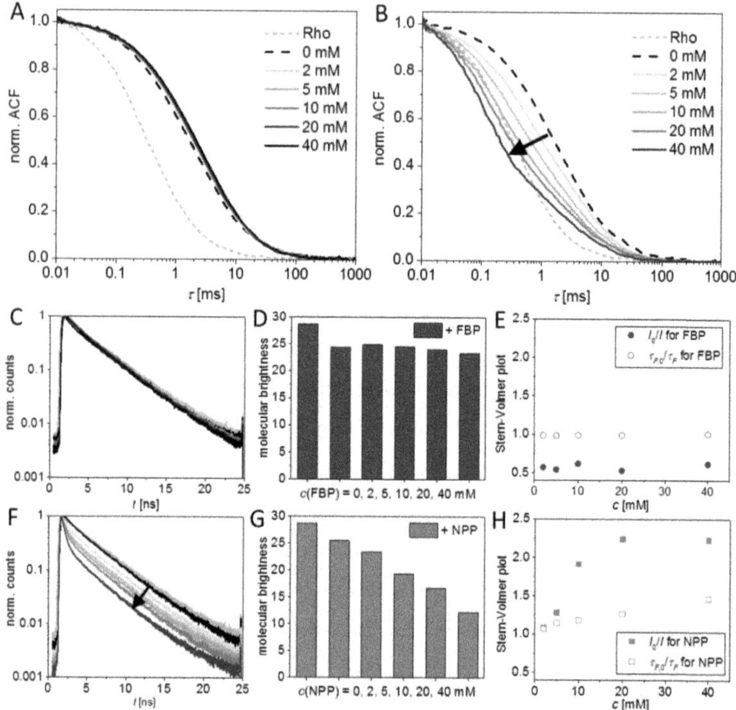

Figure 7.5. Concentration dependence of FBP (A, C-E) and NPP (B, F-H) as substrates for labeled phosphatase. Arrows indicate effect of increasing substrate concentration. A) and B) Normalized autocorrelation functions as a function of the substrate concentrations and relative to rhodamine 110 as the reference autocorrelation function. C and F) Fluorescence lifetimes measured during the experiments in A) and B) with the same color code. D) and G) Molecular brightness of the same experiments. E) and H) Stern-Volmer plot based on the average fluorescence intensity and the average fluorescence lifetime fits (using the average of the two amplitude-weighted fluorescence lifetime components for NPP).

NPP concentration (Fig. 7.5B), which shifted the FCS curves to shorter correlation times and altered the shape of its curves (similar to Fig. 7.3J) indicating increasing amounts of quenched "dark states" of the fluorophore in the 100 µs time range. The fluorescence lifetime was dominated by a strong decrease of the proportion of a long lifetime component ($\tau_F = 3.7$ ns) at higher concentrations of NPP (Fig. 7.5F). As mentioned earlier, changes of τ_F are an indicator for rate constant changes, which in turn affect τ_T as seen in the autocorrelation function (see Eq. 7.4). We calculated the molecular brightness and examined the Stern-Volmer plot of the interaction between NPP and the labeled phosphatase. As a result of quenching, the molecular brightness of the labeled phosphatase drops continuously to values below 50 % at 40 mM NPP (Fig. 7.5G). The Stern-Volmer plot revealed concentration dependent quenching both in $\bar{I}_0/\bar{I}$ and $\bar{\tau}_{F,0}/\bar{\tau}_F$, which explains the trend in the autocorrelation functions.

7.9 Conclusions

It is fascinating to contemplate that whenever an enzyme is catalytically active, it will show enhanced diffusion and possibly swim. Theory confirms that enzymes can show increased diffusion when they are catalytically active,[31,55,28,46] and that there are several factors that can contribute to the enhancement. However, there is a discrepancy between theory and experimental observations based on fluorescence correlation spectroscopy (FCS) data. We were able to theoretically describe and experimentally demonstrate effects that cause apparent increases in diffusion. In particular, we demonstrated that the apparent increase in the diffusion of F_1-ATPase in the presence of its substrate or its inhibitor[11] is largely due to the artefacts of dissociation and surface binding. In the case of alkaline phosphatase we showed that the observed dramatic change in the autocorrelation time[8] is not because of catalytically enhanced diffusion, but due to fluorescence quenching. We also found that the use of a substrate that does not cause quenching did not lead to a measurable increase in diffusion of this particular enzyme. It is now of interest to quantitatively determine how large in general the predicted increase in diffusion due to catalytic activity is and whether one can detect that enzymes "swim". Since the swimming effect seems to be close to the detection limit of conventional FCS, more advanced FCS methods are needed to quantify the exact extent of enzyme swimming.

7.10 Acknowledgment

This work was supported by the DFG in the priority program SPP 1726 as joint project 253407113 (M.B. and P.F.).

8 Label-Free Enzyme Diffusion Measurements

The manuscript "Absolute Diffusion Measurements of Active Enzyme Solutions by NMR" was published in *The Journal of Chemical Physics* in March 2019 as part of the special issue "Chemical Physics of Active Matter". As was shown in the previous chapter, several artefacts can lead to misinterpretation of enzyme FCS measurements. FCS is nevertheless a valuable tool for enzyme studies, but independent experiments were needed. The presented publication therefore utilized diffusion NMR to measure the active diffusion of fructose bisphosphate aldolase (aldolase). It is the first diffusion NMR study of an active enzyme. The use of diffusion NMR allowed measurements of unlabeled enzymes and tracer molecules. This way three claims from earlier publications could be tested:

- Claim I: Aldolase exhibits enhanced diffusion, when it cleaves fructose-1,6-bisphosphate.[9,12]
- Claim II: Aldolase exhibits enhanced diffusion upon reversible binding of its inhibitor pyrophosphate.[9]
- Claim III: Active aldolase enhances the diffusion of passive tracers in its surrounding.[67]

Within the accuracy of the diffusion NMR measurements no diffusion enhancement could be found by the author of this thesis, which therefore negates all three claims. Possible explanations for the initial results are poor sample purification (see Chapter A.1.1). Moreover, enzyme concentrations are by several orders of magnitude lower in FCS measurements compared to NMR measurements and therefore enzyme dissociation might explain the earlier FCS results. During the preparation of this publication a study supporting the negation of claim I was published by another research group that used another technique (DLS).[106]

Contributions: J.-P. G. wrote the first draft of the manuscript. All authors contributed to the finished manuscript. J.-P. G. prepared and characterized all samples and performed all the NMR measurements.

J. Chem. Phys. **150**, 124201 (2019)

Absolute Diffusion Measurements of Active Enzyme Solutions by NMR

Jan-Philipp Günther,[1,2] Günter Majer,[1,] Peer Fischer[1,2,*]*

[1]Max Planck Institute for Intelligent Systems, 70569 Stuttgart, Germany

[2]Institute of Physical Chemistry, University of Stuttgart, 70569 Stuttgart, Germany

**E-Mail: majer@is.mpg.de, fischer@is.mpg.de*

8.1 Abstract

The diffusion of enzymes is of fundamental importance for many biochemical processes. Enhanced or directed enzyme diffusion can alter the accessibility of substrates and the organization of enzymes within cells. Several studies based on fluorescence correlation spectroscopy (FCS) report enhanced diffusion of enzymes upon interaction with their substrate or inhibitor. In this context, major importance is given to the enzyme fructose-bisphosphate aldolase, for which enhanced diffusion has been reported even though the catalysed reaction is endothermic. Additionally, enhanced diffusion of tracer particles surrounding the active aldolase enzymes has been reported. These studies suggest that active enzymes can act as chemical motors that self-propel and give rise to enhanced diffusion. However, fluorescence studies of enzymes can, despite several advantages, suffer from artefacts. Here we show that the absolute diffusion coefficients of active enzyme solutions can be determined with Pulsed Field Gradient Nuclear Magnetic Resonance (PFG-NMR). The advantage of PFG-NMR is that the motion of the molecule of interest is directly observed in its native state without the need for any labelling. Further, PFG-NMR is model-free and thus yields absolute diffusion constants. Our PFG-NMR experiments of solutions containing active fructose-bisphosphate aldolase from rabbit muscle do not show any diffusion enhancement for the active enzymes nor the surrounding molecules. Additionally, we do not observe any diffusion enhancement of aldolase in the presence of its inhibitor pyrophosphate.

8.2 Introduction

Chemical reactions on the surface of microparticles can cause local concentration gradients around the particle. A local concentration gradient of substrate or product molecules can in turn give rise to fluid flows around the particle, and momentum conservation then requires that the particle moves in the opposite direction to these flows. Energy is consumed, which powers the self-propelled particles. Such "active" particles are also known as chemical motors with interesting properties and applications.[29,175–180] Whether these or related mechanisms can also influence the diffusive behavior of nanometer-sized catalytically-active materials is not yet fully understood.[31,38] Experimental diffusion studies of active molecular machines,[62] biopolymers,[1] and small nanocolloids[35] suspended in solution are challenging and lead to different results regarding the existence and magnitude of the diffusion enhancement in these nanosystems. The question whether the reaction of enzymes can cause their self-propulsion has potentially important implications for biochemistry. Recent studies suggest that the diffusion of active enzymes differs from their purely passive Brownian motion.[181,7,8,67,9,12,10] These studies rely on measurements of fluorescently labeled enzymes. However, it was recently shown that photo-physical artefacts may be misinterpreted as diffusion enhancement of active enzymes.[1] It is therefore helpful to consider alternative experimental methods that do not require fluorophores and are thus independent of any photo-physical effects. In addition, the low concentration of enzymes in fluorescence measurements can be problematic as it may cause enzyme oligomers to dissociate.

Nuclear magnetic resonance (NMR) offers unique possibilities to measure diffusion coefficients by combining pulsed field gradients (PFGs) with either a spin echo[84] or stimulated echo[83] sequences. These are often also referred to as pulsed gradient spin echo NMR (PGSE NMR), pulsed gradient stimulated echo NMR (PGSTE NMR), or, since PFG-NMR is a pseudo-2D-NMR method, as diffusion ordered NMR spectroscopy (DOSY).[182] PFG-NMR has proven to be a powerful tool to measure absolute values of the diffusion coefficients of molecules directly, *i.e.* without the need for fluorescence labeling and independent of any diffusion-model assumptions.[183–185] Unlike fluorescence correlation spectroscopy (FCS),[73] PFG-NMR does not need any calibration or reference measurement. Additionally, the diffusion constant of multiple molecules can be determined simultaneously with PFG-NMR, particularly if the molecules possess distinct NMR peaks, but this is not a necessary criterium.[184,185] We like to emphasize that in any diffusion measurement it is also possible to measure the diffusion of the solvent, which provides an internal reference and control of the accuracy of the diffusion measurement provided there is no interaction between the solvent and the molecule of interest.[186] In particular, it is possible to determine the diffusion coefficients of reactant, product[187] or solvent molecules *in situ*, which allows one to calculate the viscosity of the reaction mixture. In this study the viscosity of the reaction mixture is accessible via the diffusivity of traces of protons (mainly HDO) in D_2O using the Stokes-Einstein-Sutherland equation:[26,27]

$$D = \frac{\mathrm{k_B}T}{6\pi\eta R} \tag{8.1}$$

Here, we use PFG-NMR measurements to study the diffusive behavior of the enzyme fructose-bisphosphate aldolase (ALD) in its active state in various solutions. Enhanced diffusion of up to 35 %[12] has been reported from FCS measurements for ALD upon substrate conversion. It has also been reported that ALD can cause the enhanced diffusion of other molecules present in solution.[67] Additionally, it has been reported that ALD shows 20 % diffusion enhancement upon interaction with its inhibitor pyrophosphate (PP).[9] The theoretical explanations regarding the potential origin and magnitude of these effects differ.[46,55,28,58] PFG-NMR has been used to study the diffusion of small molecular tracers in the presence of passive proteins[186] as well as active molecular catalysts.[64] However, to the best of our knowledge, this is the first application of PFG-NMR to determine the diffusion of active enzymes.

8.2.1 PFG-NMR Principle

Using the stimulated echo sequence depicted in Fig. 8.1, the nuclei of the molecule of interest are "spin-labeled" in the *z*-direction by a magnetic field gradient G along z, which is applied for a time δ_S. The signal I acquired at the end of the stimulated echo sequence depends on the movement of the nuclei (along the *z*-direction) and the applied gradient strength. The maximum signal I_0 is measured in the absence of a gradient. If the molecule moves, its NMR signal is attenuated according to the following equation:[184,185]

$$I(G) = I_0 \cdot \exp(-b(G) \cdot D), \tag{8.2}$$

where D denotes the diffusion coefficient of the molecule and $I(G)$ is the integrated intensity of a resonance peak in the NMR spectrum. $b(G)$ is an experimental parameter that depends on the diffusion time Δ as well as the length and amplitude of the gradient pulse. If the shape of the gradient pulse is given by a half-sine function with a duration δ_S, then $\delta_G = \frac{2}{\pi}\delta_S$ where δ_G is the effective gradient time. In this case, the value $b(G)$ in Eq. (8.2) is given by[22,188,82]

$$b(G) = \gamma^2 G^2 \left[\delta_G{}^2 \left(\Delta - \frac{\pi}{8}\delta_G\right)\right]. \tag{8.3}$$

The diffusion time Δ is given by the separation of the leading edges of the gradient pulses und $\gamma = 2\pi \cdot 42.576\ \mathrm{MHz/T}$ denotes the gyromagnetic ratio of the protons for ^{1}H-PFG-NMR. In this work, all parameters except G were kept constant throughout each PFG-NMR experiment. By varying G the attenuation of $I(G)$ can be fitted with Eq. (8.2) to reveal the diffusion coefficient of the molecule (see Fig. 8.1 (b)).[184,185]

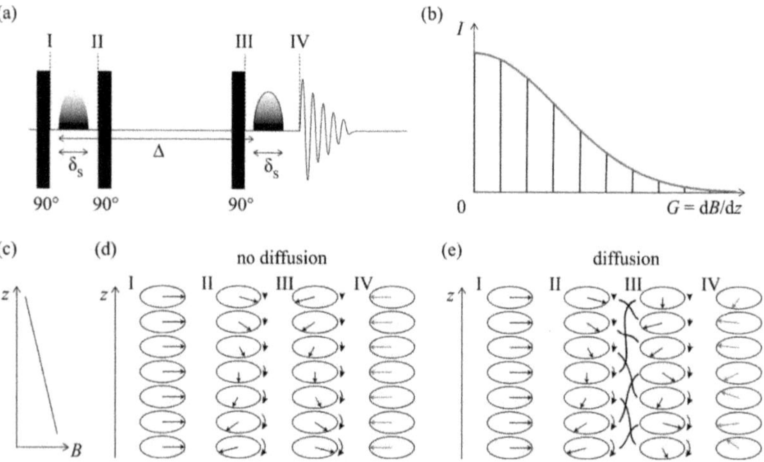

Figure 8.1. General principle of Pulsed Field Gradient NMR experiments. (a) Stimulated spin echo pulse sequence including diffusion time Δ and half-sine gradient length δ_S. (b) Dependence of the spin echo attenuation on the gradient strength G. (c) The gradient of the magnetic field B is applied along the *z*-axis of the sample. (d) Spin evolution of nuclei at different *z* positions during the NMR sequence. (e) Spin evolution during the NMR experiment including the effect of diffusion.

Proteins exhibit a wide distribution of chemical shifts in ^{1}H-NMR spectroscopy (Fig. 8.2 (a)). ^{13}C- or ^{15}N-NMR spectroscopy of proteins yields more distinct NMR spectra, but ^{1}H nuclei exhibit the highest gyromagnetic ratio and abundance in proteins. Therefore, we decided to use the ^{1}H signal together with wide signal integration for our protein diffusion studies. Accordingly, all samples have been prepared in a Tris-d_{11} D_2O buffer (Fig. 8.2 (b)), which only shows very low ^{1}H signal intensities mainly due to non-deuterated Tris (chemical shift $\delta \approx 3.5$ ppm) and traces of H_2O in D_2O (HDO, $\delta \approx 4.7$ ppm).

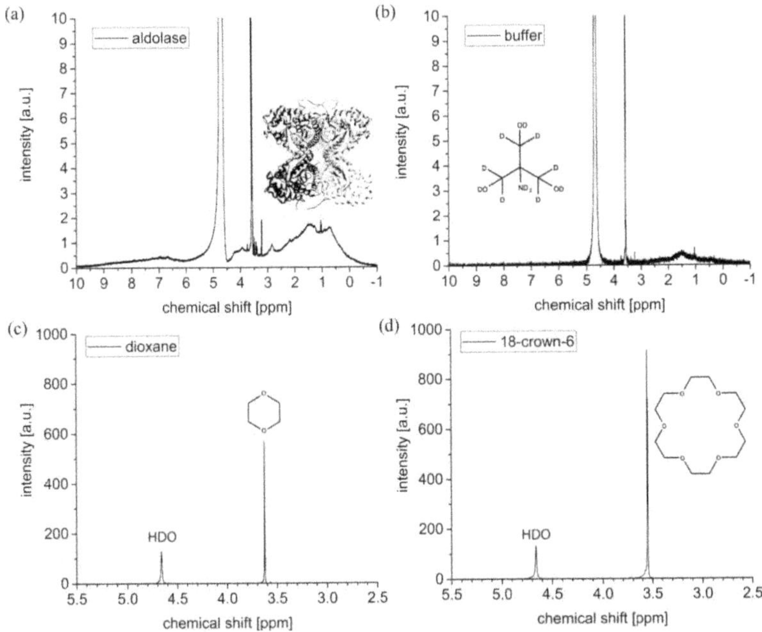

Figure 8.2. ^{1}H-NMR spectra and structures of the chemicals used in this study. (a) 11.9 µM ALD in Tris-d_{11} buffer. (b) Tris-d_{11} buffer alone. Tracer molecules dioxane (5 mg/mL) (c) and 18-crown-6 (5 mg/mL) (d) both containing residual HDO in Tris-d_{11} buffer.

Dioxane is an excellent tracer in protein ^{1}H-PFG-NMR to monitor the solution's viscosity, since it only shows one singlet peak, high solubility in H_2O and D_2O, is not charged, and does not interact with proteins.[186] To explore tracers of different sizes, we also studied 18-crown-6, which is a crown ether, a chemical analogue of dioxane, and shares all its advantages. The chemical shifts of dioxane and 18-crown-6 are similar, but spectrally well separated and therefore allow simultaneous determination of their diffusion coefficients (Fig. 8.2 (c) and (d)).

8.3 Experimental Details

8.3.1 Materials

Fructose-bisphosphate aldolase (product number A2714), triosephosphate isomerase type X (TPI) and α-glycerophosphate dehydrogenase type X (GDH) were all purchased from Sigma-Aldrich as lyophilized powders from rabbit muscle. The following chemicals D-fructose 1,6-bisphosphate (FBP) trisodium salt hydrate; anhydrous 1,4-dioxane; 18-crown-6; D_2O; 1 M tris(hydroxymethyl-d_3)amino-d_2-methane (Tris-d_{11}) solution in D_2O; 35 wt. % deuterium chloride in D_2O; and 30 wt. % sodium deuteroxide in D_2O were also purchased from Sigma-Aldrich. Anhydrous sodium pyrophosphate was purchased from Alfa Aesar.

Apart from ALD, the chemicals were used without further purification. ALD was purified as follows: ALD aggregates were removed by size-exclusion chromatography in Tris buffer using a HiPrep 16/60 Sephacryl S-300 HR column (GE Healthcare). The purified fraction of the ALD tetramer was then used in Tris-d_{11} buffer for all subsequent NMR and activity experiments. Details on buffer preparation, chromatography and buffer exchange can be found in the supplementary information (Chapter A.1.1).

8.3.2 Enzyme Activity Assay

The activity of ALD was measured in the same solution (Tris-d_{11} D_2O buffer) and at the same temperature (25 °C) as in the NMR experiments. A coupled activity assay of ALD with TPI and GDH has been performed to determine K_{FBP} and $k_{cat,f}$ of the purified ALD. The activity of ALD was observed by monitoring the consumption of NADH through its 340 nm absorption band with a CARY 4000 UV/Vis spectrometer. The Michaelis-Menten analysis of the activity data as well as additional control measurements can be found in the supplementary information (Chapter A.1.2).

8.3.3 NMR and PFG-NMR Experiments

1D-NMR spectra of molecules and the enzyme were recorded on a 400 MHz JEOL ECZ400S spectrometer in Tris-d_{11} buffer at room temperature.

Diffusion coefficients of the enzyme ALD and of the small "tracer" molecules (HDO, dioxane, 18-crown-6) were measured by PFG-NMR with a Bruker Avance III 400 MHz spectrometer. Magnetic field gradients were generated using a diff60 diffusion probe and a Great60 gradient amplifier (Bruker Biospin) with the Bruker software TopSpin 3.2. All diffusivities were measured using the pulsed gradient stimulated echo sequence with spoiler gradient pulses.[189] Gradient steps and operating parameters are listed in the supplementary information (Chapter A.1.3). In order to improve the signal-to-noise ratio, spectra were averaged up to 128 times. The diffusion probe was cooled with water to a constant temperature of 25 °C. Owing to the thermal contact, the sample temperature was stable to 25.0 °C ± 0.3 °C throughout the measurement.

The diffusion constant was extracted from the spectra by integration of the NMR intensities I and by applying a nonlinear regression to the plot of $I(G)$ with the help of a custom MATLAB code. Each experiment was repeated three times. Sample sizes were always 500 µL. The standard deviation of the data presented here was calculated via Gaussian error propagation from the standard deviation of the D extracted from the nonlinear regression. The results of the regressions of all individual experiments are listed in the supplementary information (Chapter A.1.4). PFG-NMR spectra for waterfall plots were averaged for clearer presentation over an interval of up to 0.03 ppm and then smoothed with a moving average filter of up to 8 data points.

8.4 Results

8.4.1 Diffusion Measurements of Solutions in the Presence of Active Aldolase

An active enzyme may experience enhanced diffusion due to a number of mechanisms. One possibility is that the catalytic activity causes a "mixing" of the solution. All solvent molecules and any small tracer molecules present in the solution should then also be stirred by flows induced by the enzymatic reactions. It has been reported that molecules and particles (hydrodynamic radius $R = 0.57$ nm to 1 μm) show enhanced diffusion, when active ALD is present even at low (10 nM) concentrations. These FCS-based studies found a diffusion enhancement of $D/D_0 = 1.26 \pm 0.07$ for a tracer with $R = 0.57$ nm at 1 mM FBP.[67] This cannot be explained with heating of the solution, since the catalyzed reaction is endothermic.

NMR is a powerful technique to test whether there are indeed any measurable flows due to the reaction, as the absolute diffusion coefficients of the solvent, the enzyme, and the tracer molecules can be determined. We have used three tracer molecules with $R = 0.11, 0.23$ and 0.42 nm (based on Eq. 8.1 and results from Fig. 8.3 (c)) that exhibit strong single-peak NMR spectra (see Fig. 8.2). The three tracer molecules were used to monitor the solution, when 10 nM active ALD was present, which catalyzed the cleavage of FBP to glyceraldehyde 3-phosphate (G3P) and dihydroxyacetone phosphate (DHAP):

$$\mathrm{FBP} \rightleftharpoons \mathrm{G3P} + \mathrm{DHAP} \tag{8.4}$$

The diffusion coefficients of all three molecules were monitored simultaneously and several substrate (FBP) concentrations were tested to probe whether the diffusion changes dependent on the enzyme's activity. The PFG-NMR spectrum (Fig. 8.3 (c)) shows the differences in attenuation of the signals of the different tracer molecules. The fastest diffusing molecule (HDO) exhibits strong attenuation even at low gradient field strengths visible at a chemical shift of $\delta \approx 4.7$ ppm. The peaks of dioxane and 18-crown-6 are spectrally very close to each other ($\delta \approx 3.7$ ppm and $\delta \approx 3.6$ ppm) due to their similar chemical makeup. Nevertheless, it was possible to analyze both independently, as they were spectrally resolved. Since fast repetition rates have been used, the mass concentration of dioxane was selected to be twice as much as the mass concentration of 18-crown-6 to compensate for differences in their spin-lattice relaxation. As expected, the attenuation of the stimulated echo of the smaller dioxane is stronger than for 18-crown-6.

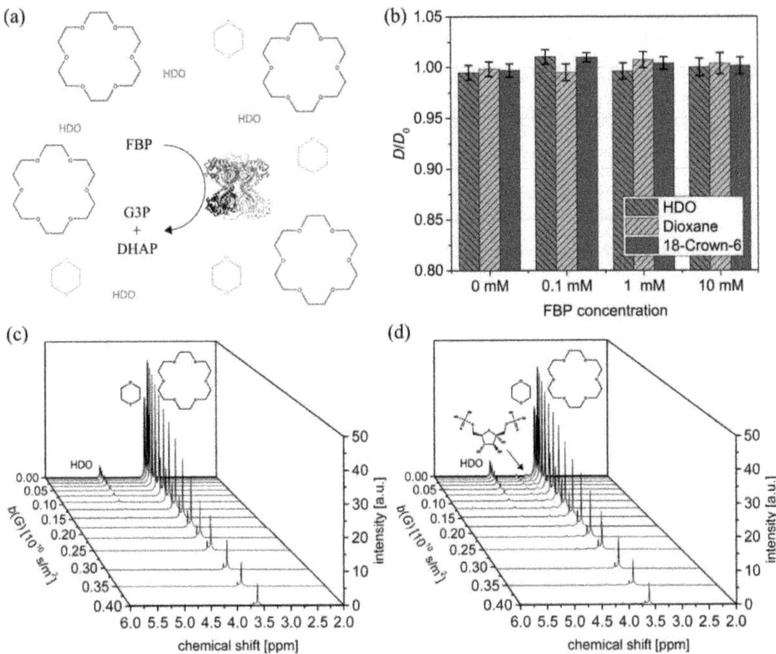

Figure 8.3. Diffusion experiments of tracer molecules in active ALD solutions. (a) Schematic representation of reaction mixture containing 5 mg/mL dioxane (green), 2.5 mg/mL 18-crown-6 (blue), 10 nM ALD (grey) and different concentrations of FBP (not to scale). (b) Fractional diffusion coefficient change (D/D_0) of tracer molecules for different substrate (FBP) concentrations (tracer diffusion D in the presence and D_0 in the absence of active ALD). Regressions in Fig. A.3 and absolute diffusion coefficients in Tab. A.3. Representative PFG-NMR spectra showing the signal attenuation of tracer molecules alone (c) and in an active ALD enzyme reaction mixture (d).

After the addition of ALD and FBP to the tracer mixture only small changes in the PFG-NMR spectrum (Fig. 8.3 (d)) are visible. Low signals of FBP can be seen around $\delta = 4\ \mathrm{ppm}$. ALD is not visible in the PFG-NMR spectrum due to its low concentration (10 nM). Since the addition of FBP has a small influence on the viscosity of the solution, we normalized the tracer diffusion coefficients of solutions containing 10 nM active ALD enzyme (D) to those without enzyme containing the same FBP concentration (D_0). D/D_0 is presented for all experimental conditions in Fig. 8.3 (b). The regressions and diffusion coefficients of each individual experiment are listed in the supplementary information (Chapter A.1.4). We found no measurable diffusion enhancement. Even for HDO and 0.1 mM FBP D/D_0 was 1.01 ± 0.01, where no diffusion change corresponds to $D/D_0 = 1$.

8.4.2 Diffusion of the Enzyme Aldolase during Interaction with its Inhibitor Pyrophosphate

Remarkably, the endothermic enzyme ALD has also been reported to show enhanced diffusion when its inhibitor is present.[9] Similar effects of diffusion enhancement without substrate conversion have recently been reported for the enzymes urease[52] and hexokinase.[190] One possible explanation is that enzymes can exhibit significant structural changes upon interaction with inhibitors and hence show enhanced diffusion. We have recently shown that even though size-changes due to inhibitor interactions occur for the enzyme ATPase, several more dominant artefacts mask this effect in FCS.[1]

Illien *et al.* argue that the observed increase in diffusion of ALD ($D/D_0 \approx 1.2$) in the presence of its inhibitor PP is due to the repeated binding-unbinding fluctuations of the inhibitor to the enzyme. This is thought to cause structural changes that significantly affect the diffusion of the enzyme ALD.[58] Since equilibrium fluctuations are persistent, they are particularly amenable to NMR studies and permit high accuracy measurements. We have therefore performed PFG-NMR measurements over several minutes to probe the diffusion of ALD in the presence of PP, which does not possess a proton and can therefore be added in high concentration.

For this study, we focused on the aliphatic range of the ALD NMR spectrum with a δ between 0.5 ppm and 1.6 ppm, since this region has the highest signal-to-noise ratio in the enzyme's NMR spectrum. In Fig. 8.4 (c) the PFG-NMR spectrum of $11.9\ \mu\text{M}$ ALD is depicted. All ALD related signals between $\delta = 0$ and 5 ppm exhibit the same attenuation, which means that all these protons are diffusing with the same diffusion coefficient. Since very high gradients have been applied, the signals of the non-deuterated Tris and remaining HDO are only visible in

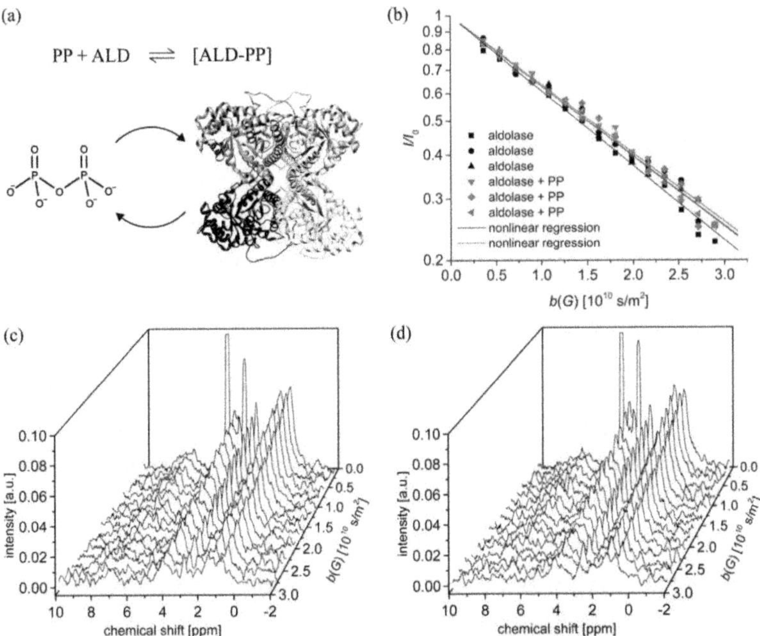

Figure 8.4. Diffusion experiment of ALD interacting with its inhibitor PP. (a) Schematic of the chemical equilibrium in the solution containing 11.9 μM ALD and 6 mM PP in Tris-d_{11} buffer (not to scale). (b) Regression of the attenuation of the aliphatic ALD signal between chemical shift $\delta = 0.5$ ppm and 1.6 ppm. The experiments were performed three times with and three times without the inhibitor PP. Representative PFG-NMR spectra showing the signal attenuation of ALD alone (c) and in the presence of its inhibitor PP (d).

the spectrum with the lowest gradient (4.242 T/m). From the PFG-NMR experiment of ALD alone, we were able to extract a diffusion coefficient D_0 for ALD of (4.7 ± 0.2) 10^{-11} m^2/s.

The diffusion of ALD was then studied in the presence of 6 mM PP, which corresponds to a 500-fold excess of inhibitor per enzyme. No additional signals could be observed in the PFG-NMR spectrum. The attenuation of the ALD signal and the corresponding regression results are shown in Fig. 8.4 (b). The averaged D/D_0 of all experiments is 0.96 ± 0.05, where D is the diffusion coefficient of ALD in the presence of PP and D_0 is the diffusion coefficient of ALD in the absence of PP. Within the accuracy of the experiment we therefore find no increase or change in the diffusion coefficient of the enzyme ALD in the presence of its inhibitor PP.

8.4.3 Diffusion of the Enzyme Aldolase during Substrate Conversion

The strongest diffusion enhancements of the enzyme ALD have been reported during conversion of its substrate FBP. Fluorescence-based measurements report diffusion enhancements of $D/D_0 \approx 1.3$.[9,12] It is quite challenging to monitor an enzyme during substrate conversion with PFG-NMR, since (pseudo-) 2D-NMR experiments are time consuming, because two parameters have to be sampled. Additionally, relatively high sample concentrations are needed especially for protein spectra. This requires large substrate concentrations, which must not deplete during the course of the measurement. We were able to obtain a satisfactory signal-to-noise ratio with a measuring time of only 4 min and 30 s, which we ran twice for each experiment. The forward turnover rate $k_{cat,f}$ of the purified ALD in Tris-d_{11} buffer used in this study is 5.73 s^{-1} ± 0.11 s^{-1} (see supplementary information (Chapter A.1.2) for details). Even though the reversible reaction reaches its equilibrium state within the first few minutes of the measurement, we estimate that > 70 % of the maximum forward reaction velocity occurs during the course of the NMR measurements (see supplementary information (Chapter A.1.2) for details), which corresponds to a turnover rate of > 4.0 s^{-1}.

The PFG-NMR spectrum of ALD recorded under the aforementioned conditions is shown in Fig. 8.5 (c). The measured diffusion coefficient D_0 is (4.6 ± 0.3) 10^{-11} m^2/s as previously determined, but now with a higher standard deviation, which is expected since a smaller number of field gradients and fewer scans per gradient are used. If the substrate FBP is present in the solution, a very strong FBP signal around $\delta = 4$ ppm can be observed, which shows a faster attenuation than the ALD peaks. However, the analysis of the ALD diffusion is not affected by FBP, since there is no spectral overlap with the ALD signal in the range between $\delta = 0.5 - 1.6$ ppm (see Fig. A.4). The ratio of the enzyme diffusion constant in the presence of substrate to the diffusion constant of the enzyme alone D/D_0 is 0.98 ± 0.09. We do not find any diffusion enhancement in our PFG-NMR measurements for active ALD.

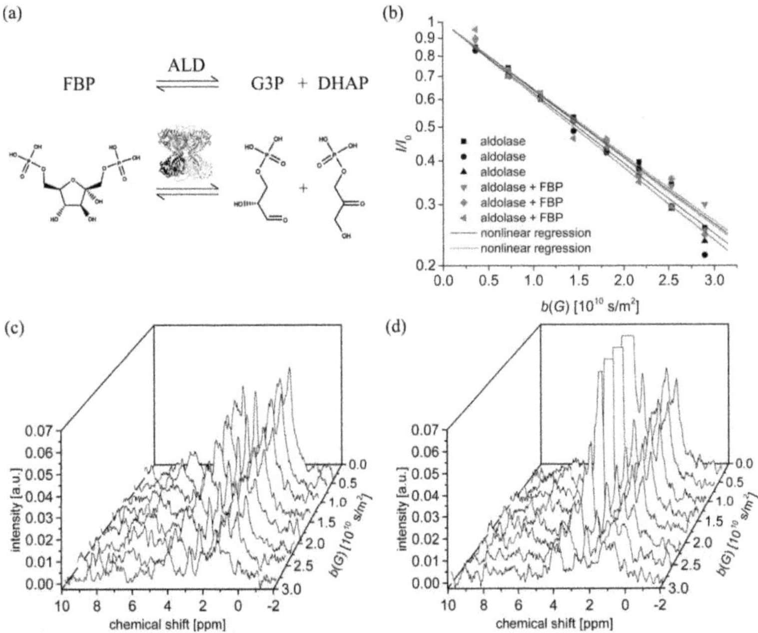

Figure 8.5. Diffusion experiment of active ALD. (a) Schematic of the reaction solution containing 11.9 µM ALD and 50 mM FBP in Tris-d_{11} buffer (not to scale). (b) Regression of the attenuation of the aliphatic ALD signal between chemical shift $\delta = 0.5$ ppm and 1.6 ppm. The experiments were performed three times with, and three times without the substrate FBP. Exemplary PFG-NMR spectra showing the signal attenuation of ALD alone (c) and during conversion of its substrate FBP (d).

8.5 Discussion

In all of our PFG-NMR measurements of the active enzyme ALD we found no enhanced diffusion. We can only speculate what may have caused other, and in particular fluorescence correlation spectroscopy (FCS) measurements, to observe enhanced diffusion.[67,9,12] We recently listed a number of mechanisms that could give rise to apparent diffusion enhancement of enzymes in FCS measurements.[1] Of special interest here is the possible dissociation of ALD under substrate conversion.[109] The low concentrations used in FCS studies may exacerbate the problem of enzyme dissociation. In Fig. A.1 (a) the chromatogram shows that many higher order aggregates are present in an ALD product from Sigma Aldrich. For this reason, we have used analytical size-exclusion chromatography to purify the ALD product prior to the measurements. The use of only molecular weight cutoff filters or short spin columns may therefore not be sufficient to exclude the presence of impurities like protein aggregates and unbound fluorophores, which may distort FCS measurements. A recent dynamic light scattering study, that also finds no diffusion enhancement of ALD during substrate conversion, uses a different, higher quality ALD product from Sigma-Aldrich,[106] which may support the hypothesis that aggregates have been present in previous FCS studies.

8.6 Conclusion

In summary, we were able to show that PFG-NMR is a useful technique to measure absolute diffusion coefficients of active enzyme solutions. Even though only results in deuterated buffer have been presented in this study, working in protonated solvents is also possible with appropriate NMR solvent suppression sequences (*e.g.* WATERGATE).[191,82] Due to the high accuracy of the PFG-NMR method even small changes in the diffusion coefficient can be resolved. This is helpful, as it is currently debated whether enzymes self-propel when they are active, *i.e.*, exhibit enhanced diffusion due to their activity. Recent fluorescence-based observations suggest enhancements of the diffusion coefficient of enzymes, when they turn over substrate and/or when an inhibitor is present. This would mean that enzymes are the smallest chemical motors. However, it has recently been shown[1] that the reported enhancements for ATPase[11] and phosphatase[8] are not due to activity, but are due to artefacts in the fluorescence measurements.

Another enzyme, for which enhanced diffusion due to activity has been reported using FCS, is ALD. ALD is a particularly interesting enzyme, because the reaction it catalyzes is endothermic, and because even interactions with its inhibitor are reported to cause increases in diffusion.[67,9,12] During the preparation of this manuscript a study utilizing dynamic light scattering has been published, where the authors find no diffusion enhancement of ALD during conversion with its substrate.[106] Our PFG-NMR study is in agreement with this latest result – we do not detect any change or increase in the diffusion of ALD when it converts its substrate (FBP). In addition, we find no measurable increase in the diffusion of the solution itself, when ALD is active and turns over substrate. We also do not find any change in the diffusion of ALD in the presence of its inhibitor (PP). Our PFG-NMR measurements therefore summarily find no enhanced diffusion when the enzyme ALD is catalytically active. It is interesting to ask what the smallest entity (molecule, enzyme or nanoparticle) is that shows enhanced diffusion and that can act as a self-propelled chemical motor. Further experiments are necessary to address this question.

8.7 Acknowledgments

We thank Henry Hess and Michael Börsch for fruitful discussions. We also thank Christine Mollenhauer and Thomas Heitkamp for support and help with protein purification. This work was supported by the DFG in the priority program SPP 1726 as project 253407113 (P.F.).

9 Enzyme Immobilization on Viruses

"Self-Assembled Phage-Based Colloids for High Localized Enzymatic Activity" was published in *ACS Nano* in March 2019. It describes the assembly and characterization of a novel kind of nanoconstruct, which consists of enzymes, viruses and inorganic particles. For this, genetically modified phages were self-assembled head-on onto magnetic microbeads. The enzyme urease was then covalently bound to the phages, which have a large surface area and should preserve the activity of urease (Chapter 6.2). Two applications of this novel construct, termed enzyme-phage-colloid (E-P-C), were then presented.

The E-P-C was used as a biocatalyst for several reaction cycles. In between the cycles, the E-P-C could easily be recovered with a magnet due to the magnetic properties of the carrier particle. An *in situ* activity assay revealed that the urease activity was increased, when the enzyme was immobilized onto the E-P-C, which might be caused by a local pH difference. In the second application, E-P-Cs were immobilized at the wall of a microcontainer with a magnet to form an enzymatic micropump (Chapter 6.3). Characterization of the pumping speeds with tracer particles revealed that this assembly strategy yielded the fastest enzymatic micropump to date.

Contributions: J.-P. G. and M. A.-C. wrote the first draft of the manuscript. All authors contributed to the finished manuscript. J.-P. G., D. R. and M. A.-C. developed and improved the enzyme-phage-colloid synthesis. J.-P. G. performed the SEM imaging and M. A.-C. the TEM imaging. J-P. G. and M. A.-C. performed the activity assay and micropump assembly and characterization. J.-P. G. analyzed the activity assay and the micropump characterization data. D. R. prepared the genetically modified bacteriophages and performed the gel-electrophoresis. J.-P. G., M. A.-C., and D. R. contributed equally.

ACS Nano **13**, 5810–5815 (2019)

Self-Assembled Phage-Based Colloids for High Localized Enzymatic Activity

Mariana Alarcón-Correa,[†,§,⊥] Jan-Philipp Günther,[†,§,⊥] Jonas Troll,[†,§] Vincent Mauricio Kadiri,[†,§] Joachim Bill,[‡] Peer Fischer,[,†,§] and Dirk Rothenstein[*,‡,⊥]*

[†]Max Planck Institute for Intelligent Systems, 70569 Stuttgart, Germany

[‡]Institute for Materials Science, University of Stuttgart, 70569 Stuttgart, Germany

[§]Institute of Physical Chemistry, University of Stuttgart, 70569 Stuttgart, Germany

**E-Mail: fischer@is.mpg.de, dirk.rothenstein@imw.uni-stuttgart.de*

[⊥]contributed equally to this work

9.1 Abstract

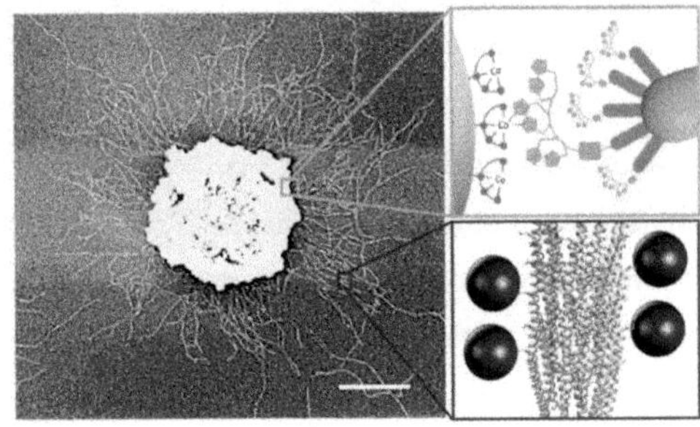

Catalytically active colloids are model systems for chemical motors and active matter. It is desirable to replace the inorganic catalysts and the toxic fuels that are often used with biocompatible enzymatic reactions. However, compared to inorganic catalysts, enzyme-coated colloids tend to exhibit less activity. Here, we show that the self-assembly of genetically engineered M13 bacteriophages that bind enzymes to magnetic beads ensures high and localized enzymatic activity. These phage-decorated colloids provide a proteinaceous environment for directed enzyme immobilization. The magnetic properties of the colloidal carrier particle permit repeated enzyme recovery from a reaction solution, while the enzymatic activity is retained. Moreover, localizing the phage-based construct with a magnetic field in a microcontainer allows the enzyme–phage–colloids to function as an enzymatic micropump, where the enzymatic reaction generates a fluid flow. This system shows the fastest fluid flow reported to date by a biocompatible enzymatic micropump. In addition, it is functional in complex media including blood, where the enzyme-driven micropump can be powered at the physiological blood-urea concentrations.

KEYWORDS: biohybrid nanostructures, phages, His tag, self-assembly, enzyme recovery, enzymatic micropumps

9.2 Introduction

The fabrication of synthetic micro- and nanoparticles, which show high activity and specificity in their catalytic activity, is challenging for conventional chemical synthesis methods. The fabrication of such sophisticated structures is necessary in order to build micro- and nanomachines and chemical motors. To date, many chemically active colloidal systems comprise components, *e.g.*, inorganic catalysts and toxic fuels, that impede their application in biological systems.[35,192] It has been demonstrated that enzymes could power synthetic biocompatible micromotors and pumps.[42,193,194] However, until now only a few enzyme-powered micropumps have been published.[24,158–160,195] One major challenge is that the immobilization of enzymes to inorganic supports often reduces the turnover rate of the enzymes, as nonbiological surfaces cause conformational changes or hinder the diffusion of the substrate.[138] Nevertheless, several papers report the immobilization of enzymes to inorganic particles and surfaces,[120,138,196,197] as well as the co-precipitation of enzymes and inorganic precursors to obtain enzymatically active aggregates.[198–201] Supports made from proteins generally represent a superior microenvironment for the immobilization of enzymes and at the same time permit their specific binding.[196] In particular, the protein surface of biological templates, such as plant viruses and phages, is advantageous for enzyme immobilization and the ensuing enzyme activity.[144,202]

In this report, we use M13 bacteriophages (phages), bacteria-infecting viruses, as biological templates for the immobilization of enzymes. Bacteriophages do not pose a threat for humans and can be genetically engineered. Genetic modifications can fuse peptides to the coat proteins of phages to facilitate the selective binding.[203] Phages have a single-stranded (ss) DNA genome encapsulated in a 6 nm wide and approximately 900 nm long proteinaceous capsid. The main structure of the viral capsid is composed of around 2700 copies of the major coat protein pVIII, which are helically arranged around the ssDNA.[204] At the ends of the filamentous structure the virus expresses five copies of the minor coat proteins pIII and pVI at one end and pVII and pIX at the other end.[205] Phages have been established as nanometer-sized building blocks serving as templates for biomineralization,[206] stress-resistant semiconductor hybrid materials,[207] piezo-active hybrid nanowires,[208] biosensing platforms,[209,210] high-power lithium ion battery electrodes[211,212] and for cell imaging.[152] These applications are possible as functional peptides can be expressed as fusion proteins on the phage surface. In addition, the major coat protein pVIII is accessible for chemical modification at three different functional groups:[213] (1) the amine group of lysine and the pVIII N-terminus,

(2) the carboxylic acid group of glutamic and aspartic acid, and (3) the tyrosine hydroxy group. Chemical functionalization is possible and has for instance been demonstrated by conjugation of dye molecules to the amine groups of the main coat protein.[152] We note that the nanometer-thin slender phage structures can be beneficial for enzyme immobilization, as the diffusion of substrate and product molecules will not be sterically hindered. Moreover, the activity of enzymes might even be higher on the phage supports compared to free enzymes.[144] Thus, phages provide a convenient scaffold for self-assembling enzymes.

Here, we present a bioinorganic hybrid construct that combines the "best of both worlds", which consists of an inorganic carrier particle for convenient manipulation and localization to which phages are coupled, which in turn serve as a proteinaceous biotemplate for enzymes. The system has a modular design and allows exploiting the advantages of each single component. The phages are genetically engineered for the expression of a molecular anchor, which attaches to the inorganic carrier particle in a self-assembly process. The magnetic property of the inorganic carrier particle allows the recycling of the anchored enzymes and their reuse in subsequent reactions. The complete assembly results in a highly active bioreactor, which can easily be moved and positioned to localize the catalytic action of the enzymes. Moreover, this enzyme–phage–colloid (E-P-C) can be used to build highly efficient, biocompatible micropumps powered by the degradation of urea. Most importantly, E-P-C-based micropumps not only are functional in artificial buffer systems but, as we show, also function in complex biological fluids like blood.

9.3 Results and Discussion

The enzyme–phage–colloids presented here consist of enzymes coupled to phages that are in turn immobilized on an inorganic support. For the fabrication of these structures, inorganic particles were first decorated in a self-assembly process with genetically modified bacteriophages. In a subsequent step, these coupled biological scaffolds were used for the immobilization of enzymes.

M13 phages were genetically modified for the expression of a His-tag fused to the N-terminus of the minor coat protein pIII (which we from now on call "3HT") (for details see Supporting Information (SI) (Chapter A.2.1)), resulting in specific binding of the phages to the magnetic particles with cobalt-based immobilization chemistry (Dynabeads, Novex), hereinafter referred to as "carrier particles". For the controlled release of the bacteriophages from the carrier particles, a protease restriction site was introduced between the His-tag and the pIII

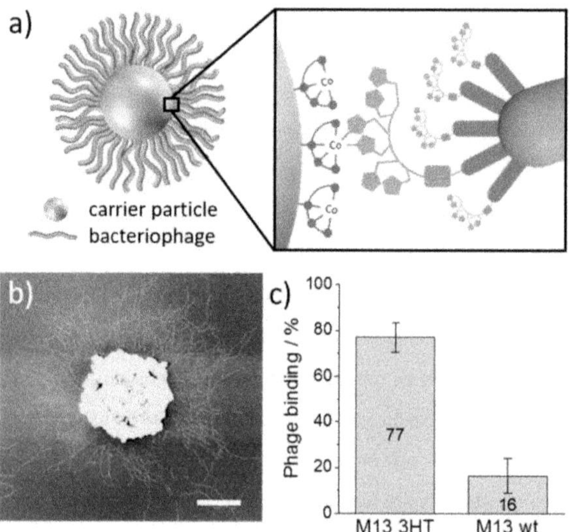

Figure 9.1. Phage–colloid (P-C). (a) Scheme of the genetically modified M13 phage 3HT (blue) expressing a His6-tag (green in inset) as fusion to the minor coat protein (blue). The His-tag is bound to tetradentate Co complexes (brown in inset) on the surface of the commercial carrier particle. The carrier particles have a diameter of 1 µm. One His-tag is schematically enlarged. The protease restriction sites are shown in red. (b) Scanning electron micrograph of a P-C structure assembled from His-tag expressing phages and a carrier particle. Scale bar: 500 nm. (c) Percentage of bound phages to the carrier particles. The phage concentration was determined by UV–vis absorption spectroscopy.

coat protein. The coupling mechanism of the phage–colloid (P-C) is depicted in Fig. 9.1a. The binding of the modified phages to the carrier particles was qualitatively determined by scanning and transmission electron microscopy (Fig. 9.1b, Chapter A.2.2), which also revealed that nonmodified phages did not bind to the carrier particles. The quantitative analysis of the P-C formation with genetically modified phages by UV–vis spectroscopy confirmed that 77 ± 7 % of the phages bind to the carrier particles. As expected, wild-type nonmodified phages (M13 wt), which do not express the His-tag anchor sequence, only showed a low nonspecific interaction with the carrier particles (Fig. 9.1c).

These P-Cs were used to bind urease enzymes (Fig. 9.2a). The modular design can in principle also be used to bind other enzymes or use other inorganic carrier particles. Urease was covalently coupled to the phages with the water-soluble cross-linker sulfosuccinimidyl (4-iodoacetyl) aminobenzoate (sulfo-SIAB, ThermoFischer Scientific). The N-hydroxysuccimide (NHS) ester of this heterobifunctional cross-linker forms a stable amide bond with the accessible amine groups of the major coat proteins of the phage. The urease is coupled to the linker via a thioether link, which is formed between the cysteine sulfhydryl groups of the urease[118] and the iodoacetyl of the linker, giving rise to the E-P-C. The amount of immobilized urease on the phage–colloid was determined by polyacrylamide gel electrophoresis (PAGE) (Fig. 9.2b). A urease calibration curve was prepared and corrected by UV–vis spectroscopy (Chapter A.2.3). The results reported in SI Tab. A.6 are based on 16 repeat experiments from seven E-P-C samples. On average 349 ± 30 phages bind to a carrier particle. We find that on average 725 ± 190 enzymes are immobilized on an E-P-C construct. This would correspond to a surface density of approximately 230 ureases/μm^2 on the inorganic carrier particle. For comparison, Patiño *et al.* immobilized 942 urease enzymes on a 2 µm SiO_2-coated polystyrene particle,[214] which corresponds to about 75 enzymes/μm^2. The catalytic activity of the bound enzymes was measured with a colorimetric activity assay (Chapter A.2.4, Fig. A.9)[123] and quantified by a coupled enzyme assay (Chapter A.2.5).[124] One major advantage of the self-assembled E-P-C is that the magnetic particle permits the recovery of all the enzymes after a catalytic process such that they can be reused (Fig. 9.2c,d). In each cycle, the E-P-Cs functionalized with urease are added to a 500 mM urea solution for 30 minutes (Fig. 9.2c,d,

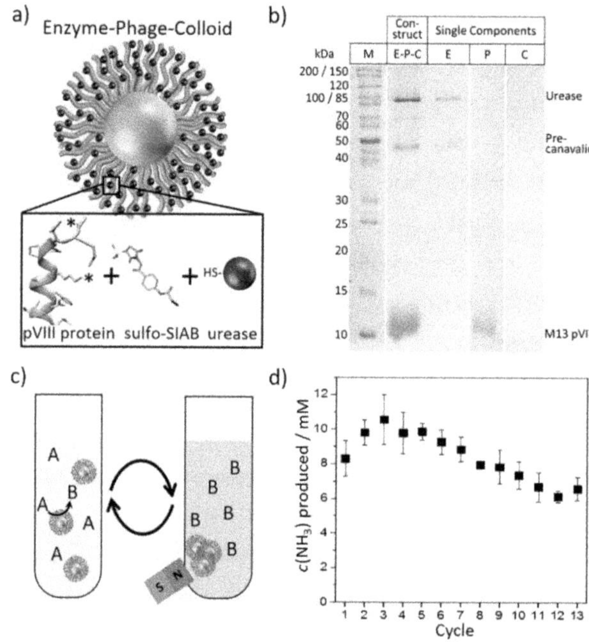

Figure 9.2. Assembly and performance of the enzyme–phage–colloid (E-P-C). (a) Schematic representation of the E-P-C. The cysteine groups of urease are coupled via the bifunctional linker sulfo-SIAB to the accessible amine groups[152] (asterisks) of the phage major coat protein pVIII (blue in inset, PDB structure from Ref. 235). (b) Polyacrylamide gel electrophoresis (PAGE) of the E-P-C construct and single components: urease enzyme (E), M13 3HT phage (P), and carrier particles (C). For better visibility, the contrast was enhanced. (c) Schematic representation of the reuse of the E-P-C for the catalytic decomposition of substrate (A) to reaction products (B) and subsequent recovery. After 30 min the E-P-Cs are recovered with a magnet and transferred to a new reaction mixture. (d) Reusability of E-P-C in subsequent reaction cycles. The E-P-C is collected by a magnet and transferred to new reaction solution. Three independent experiments are performed, and in each experiment the E-P-C is subjected to 13 subsequent cycles with a total time of more than 8 h. The activity of the enzyme was monitored by the production of NH_3. The urea is converted in all reaction cycles, indicating the time- and transfer-stability of the urease enzyme on the colloid.

Chapter A.2.4, Fig. A.10). It is seen that the constructs are stable and remain active even after 13 reaction cycles. This proves that the E-P-C constructs are a reliable and easy to use tool for catalytic applications of enzymes and could therefore be useful in the synthesis of pharmaceuticals or their precursors.[135,215]

The E-P-C constructs can also be used to realize a micropump in a microfluidic chamber. It has recently been shown that localized active enzyme patches can cause a flow in a closed container.[24,158–160,195] The whole system has been termed a "self-powered enzyme micropump".[158] The mechanism causing the convection-like behavior is thought to be driven by density differences between the enzymes' substrates and products.[159] Although thermal effects play a role, the density difference and hence the motor driving the movement analogous to convection in urease-based micropumps is primarily caused by solutal effects.[160]

We localize the E-P-Cs with a magnet at the wall of the chamber and show that the high local concentration of enzymes leads to pronounced fluid flows. The dimensions of the closed fluid chamber for the micropump system are similar to the dimensions of previously reported micropumps.[158] The pumping efficiency, reflected by the pumping speed, was determined with the help of 1 µm polystyrene tracer particles. The fluid flow is monitored at equidistant points from the center close to the chamber bottom surface and at equidistant points at a plane 550 µm above the bottom surface (Fig. 9.3a). Since the chamber is a closed system, the activity of the enzyme in the presence of the substrate will generate a concentration gradient that will cause a flow in the chamber.[159] It was found that the fluid flows outward and away from the E-P-C location (center of the chamber) on the bottom surface (in the plane of the enzymes). On the upper part of the chamber, the flow was inward (toward the center of the chamber) (Fig. 9.3a). Convection-like currents are clearly observed. The flow speeds of the tracer particles recorded in three chambers were averaged for each concentration of the substrate and at least at two different points in the chamber ("top" and "bottom"). A maximum flow speed of 6.9 ± 1.4 µm/s was observed at the top of the chamber for 10 mM urea (Fig. 9.3b). A flow decrease is observed for higher substrate concentrations. Moreover, no flow speed is observed when the urease enzymes are simply suspended in solution, and only a small flow speed is observed for urease directly coupled to the inorganic particles

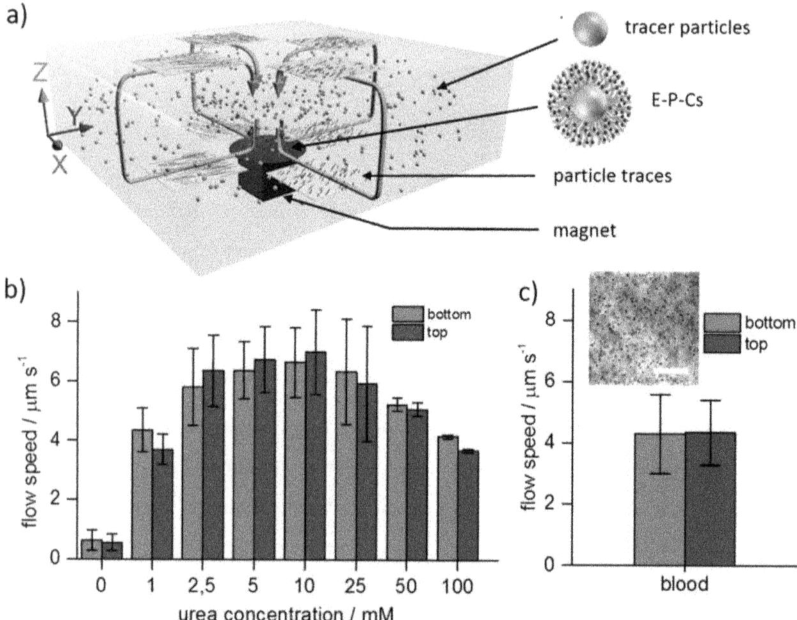

Figure 9.3. Enzymatic micropumps constructed with E-P-Cs. (a) Micropump setup: The enzyme–phage–colloids (E-P-Cs, brown) are attracted to the bottom surface of the chamber by a small permanent magnet. The chamber dimensions are (x, y, z) = (10 mm, 10 mm, 0.7 mm). The orange arrows indicate the direction of the fluid flow after the addition of urea. The insets (each corresponding to an area of 323 µm × 323 µm) show the tracked paths of tracer particles (gray) at different positions in the chamber. (b) The average flow speeds of tracer particles measured in three different micropump setups as a function of the urea concentration at the positions (x, y, z) = (0.45, 0.25, 0.15 mm) "bottom" and (0.45, 0.25, 0.55 mm) "top". (c) Average flow speeds measured in serum/blood samples. The E-P-C constructs are functional in this complex medium, where the red blood cells act as tracer particles and the urea concentration found in blood is sufficient for powering the system. Inset: Red blood cells in the micropump setup. Scale bar: 100 µm.

without phages as templates (Chapter A.2.6). We found that the E-P-C micropump system reported herein has a flow speed that is higher than previously reported micropumps utilizing other enzymes (catalase micropump).[158] Moreover, the phage-based pump is 20 times faster than other reported urease micropumps for the same concentration of urea.[158] We ascribe the efficiency to the use of the E-P-C constructs.

The urea-driven E-P-C micropump presented in this article can also be used in biological fluids. It turns out that the physiological range of urea in blood is sufficient to power the E-P-C. Blood plasma and whole blood (both from sheep) were mixed 99:1 in order to monitor the flow speed caused by the enzymatic conversion of urea naturally found in the sample. No tracer particles were added. Instead, the red blood cells were used as tracer particles. In the physiological range of urea in the plasma/blood samples (Fig. 9.3c), the flow speed was 4.3 ± 1.3 µm/s at the bottom of the cell and 4.4 ± 1.1 µm/s at the top. E-P-Cs can thus be used in blood, which suggests applications in lab-on-a-chip devices.

9.4 Conclusions

We have demonstrated that genetically modified bacteriophages can be used as scaffolds and linkers for the immobilization of enzymes. The enzyme-decorated phages can in turn be coupled to an inorganic substrate or particle. The modular approach enables flexibility in enzyme and particle combinations for specific applications. Here, urease enzymes were chemically coupled to the phages, which were bound via self-assembly to magnetic carrier particles. This enzyme–phage–colloid (E-P-C) thereby permits the localization of enzymes for heterogeneous catalysis conserving the characteristic high enzyme activity, which is often reduced when enzymes are directly bound to inorganic substrates. At the same time, the coupling of the phages to a magnetic particle allows for the convenient recovery of enzymes, as we have demonstrated. These superior properties of the E-P-C enabled the setup of an enzymatic micropump generating the fastest fluid flow observed for any enzyme-powered micropump to date. The micropump formed from the E-P-C constructs also functions autonomously in blood by converting naturally present urea. Utilizing other enzymes could allow for on-demand pumping or selective detection and quantification of substrates. Based on the modular assembly of the E-P-C, the enzyme functionalities can be exchanged and used in organic synthesis where the intrinsic chemo-, regio-, diastereo-, and enantioselectivity of enzymes is advantageous.[135]

9.5 Materials and Methods

Genetic Modification of M13 Bacteriophage Templates

M13 phages are genetically engineered for the expression of a His-tag (6 × his) fused to the N-terminus of the minor coat proteins pIII and pVII, respectively. In addition, a thrombin protease site (amino acid sequence: LVPRGS) is introduced between the His-tag and the minor coat protein in order for the controlled release of phage from the carrier particle. The genetic modifications of M13 phage templates is performed in the M13KE genome background (New England Biolabs, NEB, USA). All genetic manipulations are introduced by PCR with specific oligonucleotide primers. Detailed information is provided in the SI (Chapter A.2).

Fabrication of the Enzyme–Phage–Colloids

The modified pIII protein will bind specifically to the surface of the carrier particle (Dynabead), which is coated with a cobalt-based immobilization chemistry specifically designed to bind His-tagged proteins (Dynabeads His-Tag Isolation & Pulldown, Novex/Life Technologies, Norway). The magnetic carrier particles have a diameter of ~1 µm.

A batch of 10^{12} phages is mixed with 50 µL of carrier particles (4.4×10^{10} beads·mL^{-1}) in a binding buffer to a total volume of 700 µL. Following the protocol recommended by the manufacturer of the particles (Dynabeads, Novex), the suspension is incubated for 10 min. After incubation and washing, the phage-particles are suspended in 470 µL of phosphate-buffered saline (PBS) buffer, and 30 µL of a solution of Sulfo-SIAB (5.95 mM) is added. The reaction conditions for the bifunctional linker are based on manufacturer's recommendations (Sulfo-SIAB, Thermo Scientific, USA). Briefly, after 1 h of incubation at 20 °C in the shaker, the sample is washed using magnetic pulldown of the carrier particles and resuspended in 500 µL of PBS. A 200 µL amount of a 1 mg/mL solution of urease in PBS is mixed in, after incubation for 2 h at 10 rpm on the rotator in the dark. To quench the reaction, 3.5 µL of 1 M cysteine solution is added. The sample is washed and stored at 4 °C. Quantification of phage-carrier particles is reported in the SI.

Characterization of Enzyme–Phage–Colloids (E-P-Cs)

The number of phages bound to the colloid in the phage–colloid particles is determined by UV–vis spectroscopy according to Smith.[216] Briefly, the amount of phages in the supernatant of the phage–colloid assembly reaction is determined. Based on the input quantity the amount of bound phage was calculated.

The amount of immobilized urease enzymes on the phage–colloid is determined by PAGE. E-P-C samples and reference samples are resuspended in SDS sample buffer (DualColor protein loading buffer, Fermentas) and incubated for 5 min at 99 °C. Samples are separated on 12.5 % gels at a fixed current of 40 mA per gel. The gels were stained with Coomassi solution (PageBlue Protein Staining Solution, Thermo Scientific). The gels are photographed, and the band intensities are analyzed with ImageJ and compared to the urease calibration curve (Supporting Information (Chapter A.2.3)).

Enzyme Activity Assay

The enzyme activity of urease immobilized on phage templates is quantified using an established method.[123] Briefly, ammonia, which is produced by the catalytic decomposition of urea by urease, induces a color change of the reaction solution from yellow to blue/green.

Three stock solutions are prepared:

Solution A: 500 mM urea, 40 mM NaCl in PBS buffer at pH 7.4.

Solution B: 5 mM sodium nitroprusside, 60 mM sodium salysilate.

Solution C: 400 mM NaOH, 10 mM NaOCl.

The sample (E-P-C) and references, *i.e.*, urease standards and carrier particles, each 10 µL, are incubated for 30 min with 200 µL of solution A and then mixed with 550 µL of solution B and 400 µL of solution C. The enzyme activity was verified by the color change of the reaction solution and UV–vis measurements (Fig. A.9 and A.10).

In Situ Enzyme Activity Assay

To determine the urease activity *in situ*, a coupled optical enzyme assay is applied.[124] The enzyme glutamic acid dehydrogenase catalyzes the formation of glutamic acid from equimolar amounts of alpha-ketoglutarate and ammonia under the oxidation of the cofactor nicotinamide adenine dinucleotide (NADH). The kinetics of ammonia-producing enzymes (here: urease) can be calculated from the rate of NADH oxidation, which is spectroscopically determined. The assay is performed in 100 mM potassium phosphate buffer, pH 7.5, which is also applied for the UV–vis measurement baseline correction. The reaction solution is composed of 1.67 mM sodium alpha-ketoglutarate, 0.33 mM NADH, 0.67 units glutamate dehydrogenase, and up to 16.7 mM urea (Supporting Information (Chapter A.2.5)).

Micropump Setup

A closed fluid chamber with the dimensions of 10 mm × 10 mm × 0.7 mm is constructed with microscope glass slides and three gene frames (Thermo Scientific). The chamber is filled with 100 mM phosphate buffer, pH 7.5, and different amounts of urea substrate (or phosphate buffer for control experiments), keeping the added volume constant in all experiments at 60 µL. In order to monitor the pumping speed, 1.5 µL (25 mg·mL^{-1}) of 1 µm plain polystyrene particles (Kisker, Germany) is added as tracer particles.

A 1 mm^3 permanent magnet is positioned underneath the center of the chamber to localize the magnetic E-P-Cs, and 1–2 µL of the E-P-Cs is carefully added to the solution on top of the magnet. The chamber is closed with a glass cover slide. The magnet is removed at this stage, as the E-P-C aggregates and adheres to the glass surface.

The fluid flow is monitored at equidistant points from the center close to the chamber bottom surface and at equidistant points at a plane 550 µm above the bottom surface. The experiments are carried out on a ZEISS inverted light microscope using a 10× magnification and a Zyla sCMOS camera (Andor).

Biological Fluids Micropump Experiments

A chamber with the same conditions described in the micropump setup was used. The positioning of the E-P-Cs, as well as the monitoring of the flow was done as indicated in the previous section. Instead of filling the chamber with a buffer solution, tracer particles, and urea, we used 74 µL of plasma and whole blood from sheep (mixed 99:1) (Fiebig Nährstofftechnik, Germany). The red blood cells in the sample are used as tracer particles, and the micropump is powered by the urea naturally present in the sheep blood/serum.

9.6 Acknowledgements

The authors are grateful to J. P. Spatz, MPI for Medical Research for SEM access and J. Plitzko, P. Erdmann, and C. Liu, MPI of Biochemistry, for TEM support. This work was in part supported by the Volkswagen Stiftung under the project "Self-assembled nano-reactors via bacteriophage engineering (ROBOPHAGE)", the DFG as part of the project SPP 1726 (microswimmers, 253407113), the BW Stiftung (BioMatS-10), and the Max Planck Society.

10 Misinterpretation of Molecular NMR

The article "Boosted molecular mobility during common chemical reactions", which was published by Wang *et al.* in *Science* in July 2020,[13] contains several experimental flaws, which undermine its central conclusion that molecular catalysts transfer significant amounts of translational energy to their molecular surrounding. Therefore, a technical comment addressing those experimental flaws including an explanation for the observations in the original study has been prepared for publication and a pre-print has been deposited on the repository *ChemRxiv* in September 2020.[4] In the meantime, an updated version of this comment has been peer-reviewed and published in *Science*.[5]

Intensity changes over the course of a diffusion NMR experiment can lead to erroneous fits of the Stejskal-Tanner equation as discussed in Chapter 4.2.3. In the publication by Wang *et al.* increasing gradient strengths were used for all experiments.[13] In combination with decreasing signal intensities, which are independent of diffusion, apparent diffusion increase was reported by these authors. The diffusion increase was observed for catalyst, reactant and solvent molecules. Signal intensity loss due to consumption by the reaction can only explain the diffusion enhancement of the reactant. Another phenomenon, which effects the signal intensity of all molecular species, is the change of the spin-lattice relaxation constant T_1, when it is short compared to the recovery time in between NMR experiment repetitions. During the copper-catalyzed click reaction, which is the best studied reaction by Wang *et al.*, the Cu^{2+} concentration changes over the course of the reaction due to the reduction to Cu^{+}. This change of the concentration of paramagnetic species in the solution is causing a T_1 change[79] and hence an intensity loss over time. This appears to have been overlooked by Wang *et al.*. The comments corrects this oversight.

Contributions: J.-P. G wrote the first draft of the manuscript. All authors contributed to the finished manuscript. J.-P. G and L. L. F. performed the measurements. J.-P. G. performed the theoretical simulations. J.-P. G. and L. L. F. contributed equally.

ChemRxiv DOI: 10.26434/chemrxiv.13023164 (2020)

Comment on "Boosted molecular mobility during common chemical reactions"

Jan-Philipp Günther[1,2], Lucy L. Fillbrook[3], Thomas S. C. MacDonald[3], Günter Majer[1], William S. Price[4], Peer Fischer[*,1,2], and Jonathon E. Beves*[*,3]*

[1]Max Planck Institute for Intelligent Systems, 70569 Stuttgart, Germany

[2]Institute of Physical Chemistry, University of Stuttgart, 70569 Stuttgart, Germany

[3]School of Chemistry, UNSW Sydney, Sydney, NSW 2052, Australia

[4]School of Science, Western Sydney University, Penrith, NSW 2751, Australia

**E-Mail: fischer@is.mpg.de, j.beves@unsw.edu.au*

Abstract: *The apparent "boosted mobility" observed by nuclear magnetic resonance (NMR) diffusion measurements is the result of a known artefact. When signal intensities are changing during an NMR diffusion measurement for reasons other than diffusion, the use of monotonically increasing gradient amplitudes produces erroneous diffusion coefficient values. We show that no boosted molecular mobility is observed when shuffled gradient amplitudes are applied.*

In "Boosted molecular mobility during common chemical reactions"[13] the authors use diffusion nuclear magnetic resonance (NMR)[22] to show several chemical species, including solvents, experience higher diffusivities during chemical reactions. The authors assign this increased diffusion to transduction from chemical into mechanical energy.[13] Similar claims of enhanced self-diffusion have been reported with enzymatic reactions,[217,218] but some of the experimental evidence has been recently questioned.[1,2,106,219] A report of diffusion enhancement of small molecules during Grubbs metathesis[64] was later shown to be caused by convection artefacts in the NMR measurements.[65] The authors of Ref. 13 took care to avoid such convection artefacts in their measurements, but did not fully account for non-diffusion-based changes in signal intensity during the reaction. This oversight is apparent from the inconsistency in reported diffusion values of the reactant propargyl alcohol. The ^1H NMR signal of the alkyne proton at 2.6–2.8 ppm and of the methylene protons at 4.1 ppm of the same molecule are reported to have diffusion coefficients of 8.55×10^{-10} m^2s^{-1} and 6.62×10^{-10} m^2s^{-1}, respectively (Tab. S4 in Ref. 13). This would imply that different locations within the same small molecule possess diffusion coefficients differing by almost 30 %.

As two independent laboratories, we have reproduced the copper-catalyzed click chemistry reaction as presented in Fig. 1 in Ref. 13 as this reaction showed the largest changes and has the most data supplied. We used very similar diffusion NMR parameters (*e.g.* double stimulated echo and stimulated echo, 2.0-2.5 ms effective gradient pulse length, 25-50 ms diffusion time, 47-49 G cm^{-1} maximum magnetic field gradient strength, and 16 gradients in linear and square root distribution). Under the conditions reported in [13] the signal intensities significantly changed during the diffusion measurements (*e.g.* over 5 min for 16 gradients), and these changes completely account for the observed changes in apparent diffusion coefficients.

The known artefact[86] occurs when significant signal intensity changes occur *during* the diffusion measurement, which leads to a systematic error in the regression of the Stejskal-Tanner equation:[22]

$$I = I_0 \exp(-b(G)D), \tag{10.1}$$

where I is the echo intensity, I_0 the echo intensity in the absence of gradient pulses, b the diffusion weighting factor, which is a function of the gradient pulse strength G, and D the diffusion coefficient. Typically, NMR diffusion measurements are performed using a series of

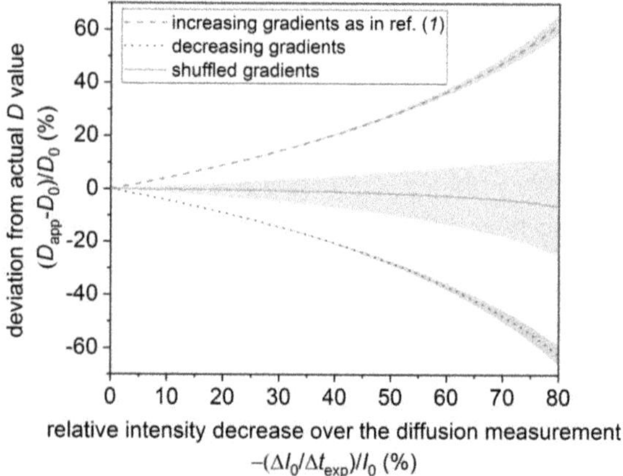

Figure 10.1. Simulated data to show the effect of changing signal intensities during a diffusion NMR experiment. D_{app} = apparent diffusion coefficient, D_0 = diffusion coefficient in the absence of the reaction, ΔI_0 = change in signal intensity, Δt_{exp} = time required for diffusion measurement. Semi-transparent areas represent the standard deviation of the fit.

monotonically increasing gradients G. If the signal intensity I_0 is also a function of time t, then changes in G will also be correlated with changes in I_0. This can result in excellent data fits, but with incorrect D values[86] as can be seen in Ref. 13. The magnitude of the artefact is determined by the magnitude of signal intensity changes during one diffusion measurement. That is, longer measurements or faster changes in signal intensity will result in data fits with greater deviations from actual diffusion coefficients (dashed red line in Fig. 10.1). This effect is reversed when decreasing instead of increasing gradients are used (dotted blue line in Fig. 10.1). Shuffled gradient amplitudes[86,220,221] can be used to estimate the real diffusion coefficient in presence of a dynamic signal intensity change (solid green line in Fig. 10.1).

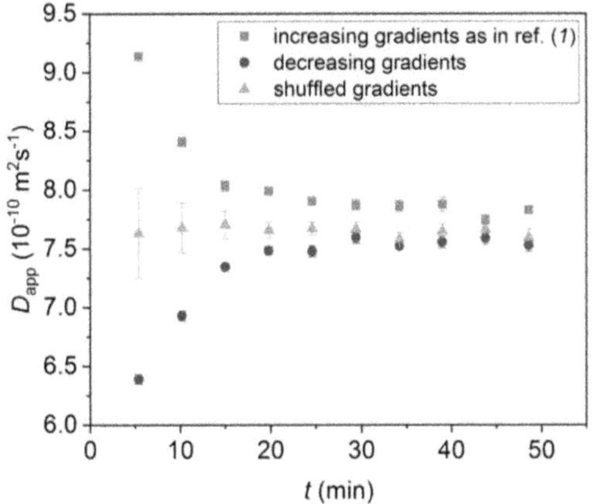

Figure 10.2. Measured diffusion coefficients for the reactant azidoacetic acid (3.9 ppm) during the click reaction. Parameters used: concentrations as in Fig. 1 of Ref. 13, double stimulated echo *diffDste*, 2.5 ms effective gradient pulse length, 25 ms diffusion time, 47 G cm^{-1} maximum magnetic field gradient strength, 16 gradients in square root distribution, linear regression as in Ref. 13.

When a monotonically increasing gradient order is used, as in Ref. 13, we also observed apparent diffusion enhancement (red in Fig. 10.2) of the reactant's signal, which decayed over time, as reported in Ref. 13. However, the opposite effect—*slower diffusion*—is observed when the gradients are ordered in decreasing strength (blue in Fig. 10.2). The authors of Ref. 13 have compensated concentration changes with "0.2% at each magnetic field gradient". Such a small intensity compensation (less than 4% over the course of the diffusion experiment) cannot account for the observed apparent diffusion change in Fig. 10.2. This indicates that the measured diffusion enhancement must be an artefact, as the true diffusion coefficient cannot depend on the order of the applied gradients. The signal intensities are not just a function of concentration, but also depend on spin relaxation.[79] We find the spin-lattice relaxation constant T_1 for all species including the solvent changes during the reaction. For example, the T_1 for azidoacetic acid changes from ~50 ms to over 2 s within the first 20 min of the reaction. This dramatic increase in T_1 fully accounts for the intensity change and hence the apparent diffusion increase. The artefact vanishes when a shuffled gradient order is used (green in Fig. 10.2), as previously shown to be required for time-resolved diffusion NMR measurements of dynamic systems.[86,220,221]

We demonstrate that there is no enhanced or "boosted" diffusion during the click reaction in Ref. 13, shown here for the reactant azidoacetic acid. Changes in signal intensities unrelated to diffusive attenuation, including for the residual solvent signals, are the source of the artefact, which results in apparent changes in measured diffusion coefficients. Since the authors of Ref. 13 state that they used increasing gradients for all diffusion NMR measurements, the artefact presented here applies also to the remaining reactions of Ref. 13.

11 Conclusions

11.1 Active Diffusion of Enzymes and Chemical Catalysts

Precise diffusion measurements of catalytically active enzymes and molecular catalysts are important for a fundamental understanding of life and chemical reactions. This thesis presents advanced diffusion measurement techniques, which challenge the interpretation of a number of recent reports in the literature that have (erroneously) claimed that enzymes and molecular catalysts are active matter, in particular Refs. 8,67,9,12,13.

The active enzyme diffusion hypothesis is mainly supported by FCS measurements. Artefacts in these measurements can therefore lead to misinterpretation of enzyme FCS data and therefore could falsify the active enzyme diffusion hypothesis. Within this thesis (Chapter 7 and Ref. 1), four possible artefacts were presented that can arise in enzyme FCS measurements and can lead to the incorrect conclusion that enzymes show enhanced diffusion: Oligomer dissociation, surface binding, conformational change, and fluorophore quenching. All four artefacts were supported with theoretical estimates and experimental examples. Prior to the work of this thesis, alkaline phosphatase was believed to be the enzyme that showed the highest diffusion enhancement.[8] However, as part of the work presented herein, it was possible to falsify these results. Careful FCCS measurements could show that fluorophore quenching by the substrate NPP is the cause for the reported misinterpretation. Careful measurements with a non-quenching substrate could then completely rule out enhanced diffusion for alkaline phosphatase. The publication discussed in Chapter 7 (Ref. 1) therefore presents the first experimental study to question the reports in the field of active enzyme diffusion. Later, other research groups could reproduce measurements from Chapter 7 (Ref. 1) and verify several of its hypotheses using independent measurement techniques.[106,222,107,219] The authors of Ref. 219, for instance, came to the same conclusions that alkaline phosphatase does not exhibit enhanced diffusion and that the initial misinterpretation of FCS data was based on fluorescence quenching by NPP. The additional use of single particle tracking (SPT) and anti-Brownian electrophoretic (ABEL) traps by the authors of Ref. 219 further substantiates the conclusions of Chapter 7 (Ref. 1). Fluorescence quenching was also identified as an artefact in acid phosphatase FCS experiments.[222]

Aldolase is an important enzyme in the context of this thesis, as it was the first enzyme reported to exhibit enhanced diffusion and catalyze an endothermic reaction.[9] Aldolase therefore challenged early theoretical explanations of the phenomenon.[8] Three claims were initially reported on aldolase regarding enhanced diffusion: I) Aldolase exhibits enhanced diffusion during (endothermic) catalysis.[9,12] II) Aldolase exhibits enhanced diffusion during reversible binding of its inhibitor.[9] III) Aldolase transfers kinetic energy to passive tracer molecules and particles in its surrounding.[67] Claim II) is particularly unphysical as binding/unbound-fluctuations do not release any Gibbs free energy, which could be used to propel the enzyme. Nevertheless, diffusion enhancement of this enzyme has been reported by others who undertook FCS measurements.[9] It has been speculated already in Chapter 7 (Ref. 1) that the apparent diffusion enhancement reported for aldolase could be a misinterpretation of aldolase dissociation into its subunits initiated by the presence of its substrate. The smaller subunits would then naturally diffuse faster independent of their catalytically activity due to their smaller hydrodynamic radius. This can be rationalized by the increase of the dissociation constant of aldolase subunits in the presence of its substrate (or inhibitor). The degree of dissociation would then be a function of substrate concentration and would be most pronounced for measurements with low enzyme concentrations. In Chapter 8 (Ref. 2) the claims I)-III) have been tested with diffusion NMR, which allowed label-free diffusion measurements of aldolase under various conditions and simultaneous measurements of tracer molecules including the solvent. The publication on which Chapter 8 (Ref. 2) is based presents the first diffusion NMR study of an active enzyme. The diffusion NMR measurements, which are performed at high enzyme concentrations, did not find diffusion enhancement in either of the conditions I)-III). These findings on claim I) are supported by DLS measurements of another group,[106] which were performed at similar enzyme concentrations as the NMR measurements in this thesis. This further supports the dissociation hypothesis for aldolase, since dissociation predominately effects measurements at low enzyme concentrations. Additionally, aldolase aggregates, which were observed in the commercial product (see Chapter A.1.1) and which were not removed in studies by others,[9,107] could have interfered with FCS measurements of other groups. The clean-up in these reports[9,107] was found to be insufficient, as the used molecular weight cut-off filters cannot remove the aforementioned aggregates. It was later shown with DLS measurements that aldolase indeed dissociates at nanomolar concentrations after 20 min and that this completely accounts for the previous (incorrect) reports of enhanced diffusion with FCS.[107] Therefore, aldolase can be ruled out as a contender for enhanced diffusion, which significantly advances the field.

Dissociation was also reported to be the cause of diffusion enhancement for several other enzymes. For instance, the hexokinase dimer dissociation constant changes in the presence of its substrate,[223,224] which was predicted[106] and later shown[107] to be the cause of the observed diffusion enhancement. Due to the large conformational changes of hexokinase during its catalytic cycle (Chapter 7, Ref. 1), hexokinase has been the most promising candidate for those who proposed that enzymes can swim because of (non-reciprocal) shape-changes during its catalytic cycle. Additionally, F_1-ATPase was shown within this thesis to dissociate in presence of its substrate or inhibitor, which explains earlier observations[11] of diffusion enhancement (Chapter 7, Ref. 1).

Urease is the enzyme for which most reports have appeared that claim diffusion enhancement (see Tab. 11.1) and it has thus been called the "gold standard"[107] for the active enzyme diffusion hypothesis. However, urease was also found to dissociate under certain experimental conditions.[107] In contrast to aldolase and hexokinase, it was claimed that urease nevertheless shows active diffusion enhancement of ca. 10 % at low substrate concentrations without dissociation. The authors in Ref. 107 justify this claim by comparing measurements, which directly measure size (static light scattering (SLS) and size-exclusion chromatography (SEC)), with measurements which directly measure diffusion (DLS and FCS). Below 1 mM urea, no dissociation was reported to occur in SLS and SEC measurements for urease, but up to 10 % diffusion enhancement were reported with FCS and DLS.[107] These 10 % diffusion enhancement are closer, but still above the theoretically expected thermodynamic limit of diffusion enhancement based on the Gibbs free energy released by the urease reaction.[14] The

Table 11.1 Self-phoresis reports for enzymes measured with self-diffusion techniques. D/D_0 denotes the maximum ratio of translational diffusion constant with (D) and without (D_0) interacting substrate/inhibitor. o.e. = other effect explains diffusion enhancement, n.r. = null result, ‡ was extracted from corresponding figure.

substrate or [inactive/competitive inhibitor]	method	D/D_0	first author	Ref.
	F_1-ATPase (*Escherichia coli*)			
adenosine triphosphate	FCS	1.23	Börsch	11
[imidoadenosine-5'-triphosphate]	FCS	1.14	Börsch	11
adenosine triphosphate	**FCCS**	**o.e.**	**Günther**	**1**
	urease (*Canavalia ensiformis*)			
urea	FCS	1.28	Muddana	7
urea	FCS	1.3‡	Riedel	8
urea	STED-FCS	1.5‡	Jee	10
urea	STED-FCS	1.8	Jee	52
[acetohydroxamic acid]	STED-FCS	1.4‡	Jee	52
[boric acid]	STED-FCS	1.5‡	Jee	52
urea	FCS	1.24	Ghosh	222
urea	SPT	3	Xu	89
urea	STED-FCS, DLS	1.1, o.e.	Jee	107
	catalase (*Bos taurus*)			
hydrogen peroxide	FCS	1.45	Sengupta	23
hydrogen peroxide	FCS	1.3‡	Riedel	8
hydrogen peroxide	FCS	1.5‡	Jiang & Santiago	51
	alkaline phosphatase (bovine)			
p-nitrophenyl phosphate	FCS	1.8‡	Riedel	8
p-nitrophenyl phosphate	**FCCS**	**o.e.**	**Günther**	**1**
fructose-1,6-bisphosphate	**FCCS**	**n.r.**	**Günther**	**1**
p-nitrophenyl phosphate	ABEL, SPT, FCS	n.r., o.e.	Chen	219
	triose phosphate isomerase (*Saccharomyces cerevisiae*)			
D-glyceraldehyde 3-phosphate	FCS	n.r.	Riedel	8
	fructose bisphosphate aldolase (rabbit)			
fructose-1,6-bisphosphate	FCS	1.3‡	Illien & Zhao	9
[pyrophosphate]	FCS	1.2‡	Illien & Zhao	9
fructose-1,6-bisphosphate	FCS	1.35	Zhao & Palacci	12
fructose-1,6-bisphosphate	DLS	n.r.	Zhang	106
fructose-1,6-bisphosphate	**NMR**	**n.r.**	**Günther**	**2**
[pyrophosphate]	**NMR**	**n.r.**	**Günther**	**2**
fructose-1,6-bisphosphate	STED-FCS, DLS	o.e.	Jee	107
	acetylcholinesterase (*Electrophorus electricus*)			
acetylcholine	STED-FCS	1.2‡	Jee	10
acetylcholine	DLS	1.2‡	Jee	107
	hexokinase (*Saccharomyces cerevisiae*)			
D-glucose, adenosine triphosphate	FCS	1.38	Zhao & Palacci	12
D-glucose, adenosine triphosphate	FCS	1.46	Mohajerani & Zhao	190
[D-glucose]	FCS	1.31	Mohajerani & Zhao	190
[adenosine triphosphate]	FCS	1.2‡	Mohajerani & Zhao	190
D-glucose, adenosine triphosphate	DLS	o.e.	Jee	107
	Na+/K+-ATPase (porcine)			
adenosine triphosphate	FCS	1.37	Ghosh	222
	acid phosphatase (potato)			
p-nitrophenyl phosphate	FCS	o.e.	Ghosh	222
adenosine triphosphate	FCS	1.35	Ghosh	222

authors of Ref. 107 further speculate that enzyme dissociation at substrate concentrations greater than the Michaelis constant k_M is a common biological feature of enzymes, but this hypothesis is highly speculative and needs further testing. Additionally, there is still no explanation for the 200 % diffusion enhancements observed for urease with SPT in highly viscous environments, especially since dissociation was ruled out for these experiments.[89] A similar mechanism to the one utilized by the bacterium *Helicobacter pylori*, which uses urease to change the pH and therefore viscosity of its surroundings, has been suggested,[217] but so far no experimental evidence for viscosity changes have been reported.

Prior to the publication of the results in Chapter 7 (Ref. 1), seven out of eight tested enzymes were reported to show enhanced diffusion (see Tab. 7.1). The theoretical and experimental demonstration of the artefacts specific to enzyme FCS measurements discussed in this thesis and in particular Chapter 7 (Ref. 1) showed errors in the reported measurements. This thesis therefore shows mistakes in five publications[11,8,67,9,12] and thus corrects the scientific record. As shown in Tab. 11.1, to date ten enzymes have been studied. For five of these enzymes active diffusion has been ruled out entirely and several studies[106,222,107,219] have now confirmed the conclusions and predictions presented in Chapter 7 (Ref. 1). Whether the remaining enzymes withstand thorough experimental investigations remains to be seen. Theoretical discussions vary in their predictions[14,28,52,55,58,59] and not all of the possible measurement artefacts that are discussed in Chapter 7 (Ref. 1) have been tested for these remaining enzyme systems. Therefore, high precision independent control experiments are still needed to also examine the remaining enzyme systems.

As has been discussed in Chapter 3.3 and 10 (Ref. 4), similar claims of enhanced diffusion as for enzymes have been made for molecular catalysts. Some of the reported measurements were shown to suffer from convection artefacts that have been misinterpreted as enhanced molecular diffusion.[63–65] However, a very recent study published in the journal *Science* claimed that enhanced diffusion of molecular catalysts and their surrounding molecules is a general phenomenon in chemical reactions.[13] The authors of Ref. 13 ruled out convection, as their measurements are based on double stimulated echo sequences. However, the author of this thesis could, together with colleagues, show that also this NMR study[13] contains several experimental flaws, which have been discussed in Chapter 10 (Ref. 4). A decrease in the intensity over the course of a diffusion NMR experiment can lead to the incorrect conclusion that there is diffusion enhancement of the catalyst, substrate, and solvent molecules. In the case of Ref. 13, however, the intensity change is primarily caused by changes in the spin-lattice relaxation, as the reactions involve paramagnetic species in solution whose concentration changes during the course of the reaction. This appears to have been overlooked by the authors of the *Science* study. In this thesis work, this interpretation is supported by theoretical simulations and experiments with deliberately changed pulsed-field gradient order, which reverses or cancels the artefact depending on the order (Chapter 10 and Ref. 4). The work in this thesis therefore calls into question the far-reaching claims of Ref. 13 on "Boosted molecular mobility during common chemical reactions", which would have significant impact and change our general understanding of chemical reactions.

11.2 Enzyme-Phage-Colloids

So far this thesis has carefully examined the claims by others of enhanced diffusion of enzymes and molecules that undergo reactions. The main finding is that these hydrodynamic interactions, which would lead to enhanced diffusion or "swimming" of individual molecules, are absent. Nevertheless, enzymes can collectively cause hydrodynamic effects, which are due to a different mechanism – one that has been discussed in Chapter 6 and demonstrated in Chapter 9 (Ref. 3).

A promising approach to utilize enzymes for this purpose is in biohybrid systems, which combine the advantages of inorganic and biological materials. Whereas immobilization of enzymes on inorganic surfaces can lower their activity, such effects are less likely for organic templates. For this reason, filamentous viruses, namely M13 bacteriophages (phages), were used in this thesis to immobilize enzymes (see Chapter 9 and Ref. 3). Genetically modified phages, which self-assembled onto magnetic carrier particles, were used in this thesis to dramatically increase the available surface area for enzyme immobilization. Urease was then covalently bound to the phage instead of directly to the carrier particle. The novel construct was termed enzyme-phage-colloid (E-P-C). E-P-Cs were shown to have higher enzyme surface loading and activity compared to direct immobilization of the enzyme onto the carrier particle. Due to the magnetic properties of the carrier particle, E-P-Cs can be used repeatedly for biocatalysis with simple magnetic recovery. It is also possible to easily assemble and power enzymatic micropumps with E-P-Cs. In enzymatic micropumps hydrodynamic flows are caused by buoyancy effects (Chapter 6.3). Enzymatic micropumps constructed from E-P-Cs were shown within this thesis to exhibit the fastest flow speeds of an enzymatic micropump to date. Another advantage of the E-P-C is that urease can readily be replaced by another enzyme, which might lead to other interesting applications. It could already be shown that E-P-Cs with urease can pump blood at physiological urea concentrations (Chapter 9, Ref. 3), which might enable lab-on-a-chip applications. The same enzyme immobilization strategy on phages was later shown to enable an enzymatic flow-through reactor based on phage nanonets.[225]

11.3 Outlook

In his famous lecture "There's plenty of Room at the Bottom",[226] Richard Feynman did not consider "active matter", even though protein motors and many of life's machinery operate out of equilibrium. One of the most fascinating claims recently is that not only protein motors can actively move, but that many enzymes and molecular catalysts also show enhanced diffusion when they are catalytically active. This thesis has carefully examined these claims and could show mistakes in the measurements by others,[11,8,67,9,12,13] which questions these claims. Of the enzyme and molecular systems examined the author of this thesis thus far, finds no evidence for any enhanced diffusion of catalytically-active enzymes and molecules.

Since it is clear that larger catalytically-active microparticles swim, it is advisable that enhanced diffusion studies should therefore rather focus on larger self-assembled structures, where rotational diffusion is comparatively slower. As has been discussed in Chapter 3.1, self-propelled nanostructures down to 15 nm have been realized with inorganic catalysts. An increase in their diffusion seems to be possible due to the higher catalytic turnover per surface area of noble metal catalysts. Therefore it is interesting to contemplate, if larger (>50 nm), but organic nanostructures could propel themselves. Besides an academic interest this might be useful for medical applications, where biocompatibility and –degradability are an issue. Furthermore chemotaxis towards a substrate of an enzyme could be used for targeted drug delivery. Suitable templates for such organic swimmers could be DNA origamis or bacteriophages (see Chapter 6.2). Both have the advantage of programmable surface functionalities and might yield structures that are big enough to support self-propulsion, when functionalized with a sufficient number of enzymes.

To return to Feynman's speculations regarding artificial nanomachinery, one could examine the potential of molecular motors such as F_oF_1-ATP synthase to propel artificial nanostructures, as this might be another fascinating way to harvest the potential of motor proteins for man-made structures.

A Appendix

A.1 Enzyme Diffusion NMR Supplementary Information

This appendix is reprinted with minor formatting adjustments with permission of AIP Publishing from the Supplementary Information of Ref. 2 (Chapter 8).

A.1.1 Aldolase Purification

Fructose-bisphosphate aldolase (ALD) was dissolved in Tris buffer (150 mM Tris, 100 mM NaCl, pH 7.4, in H_2O) at a concentration of 5 mg/mL or higher. Up to 5 mL of this solution were injected onto a HiPrep 16/60 Sephacryl S-300 HR column (GE Healthcare), which previously has been flushed with Tris buffer. The flow speed during the purification was 0.5 mL/min and the ALD tetramer was collected typically in the elution volume between 64 and 72 mL (Fig. A.1 (a)). Purified ALD was then concentrated with Amicon Ultra-4 10k (Merck Millipore) molecular weight cutoff filters and the buffer was changed to a Tris-d_{11} buffer with Micro Bio-Spin 6 (Bio-Rad) spin columns. The Tris-d_{11} buffer contained 100 mM Tris-d_{11} in D_2O, which was adjusted with a pH electrode and DCl and NaOD to a pD of 7.4.[227] The concentration of the ALD solution in Tris-d_{11} buffer was then determined via UV spectroscopy

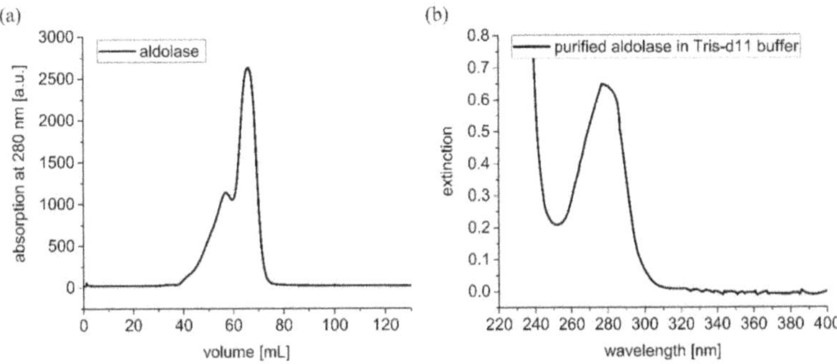

Figure A.1. Purification of ALD. (a) Chromatogram of ALD on a Sephacryl size exclusion column in Tris buffer. (b) UV extinction spectrum of aldolase tetramer fraction (64-72 mL) after buffer exchange to a Tris-d_{11} buffer.

(Fig. A.1 (b)) using the theoretical extinction coefficient of the ALD tetramer at 280 nm of $\varepsilon = 134{,}100\ (\mathrm{M} \cdot \mathrm{cm})^{-1}$ (calculated[228] from the PDB entry 1ZAH[113]).

A.1.2 Aldolase Activity Assay

The ALD activity was determined in a coupled cascade assay of ALD, triosephosphate isomerase (TPI) and α-glycerophosphate dehydrogenase (GDH) by monitoring the consumption of NADH at 340 nm.[112] D-Fructose 1,6-bisphosphate (FBP) is cleaved by ALD to glyceraldehyde 3-phosphate (G3P) and dihydroxyacetone phosphate (DHAP) (see Eq. (A.1)). TPI then converts G3P through isomerization into additional DHAP (see Eq. (A.2)). The final reaction of the cascade is the reduction of DHAP to glycerophosphate, which is catalyzed by GDH (see Eq. (A.3)). This consumes NADH, which leads to a diminishing UV extinction signal at 340 nm. The activity of ALD can be calculated from the slope of the UV signal of NADH over time. Two NADH are consumed per cleaved FBP. Typically 33-200 nM ALD, 1.3 nM TPI, 59 nM GDH and 233 µM NADH were mixed with concentrations of FBP between 0 and 50 mM. The whole assay was performed in Tris-d_{11} buffer to be comparable to the NMR experiments. The temperature during the activity assay was 24-25 °C. Three measurements were performed for each condition tested.

FBP $\xrightleftharpoons{\text{ALD}}$ G3P + DHAP (A.1)

G3P $\xrightleftharpoons{\text{TPI}}$ DHAP (A.2)

$$+ \ \mathrm{NADH} + \mathrm{H}^+ \ \xrightleftharpoons{\mathrm{GDH}} \ + \ \mathrm{NAD}^+ \tag{A.3}$$

The conversion of FBP to DHAP and G3P by ALD is reversible. This is expressed in the rate equation Eq. (A.4),[229] where v is the net conversion rate of FBP, $v_{max,f}$ the maximum forward reaction velocity, $v_{max,r}$ the maximum reverse reaction velocity and K_{FBP}, K_{G3P} and K_{DHAP} are the Michaelis constants of the respective molecules. Since the products DHAP and G3P are quickly converted in the kinetic assay by TPI and GDH, Eq. (A.4) can be simplified using the assumption that $[G3P] = [DHAP] = 0$. This simplifies the rate equation to a standard Michaelis-Menten equation (Eq. (A.5)), which has been fitted to the data of the kinetic assay (Fig. A.2). This leads to a $k_{cat,f} = v_{max,f}/[ALD]$ of $5.73\ \mathrm{s}^{-1} \pm 0.11\ \mathrm{s}^{-1}$ and a K_{FBP} of $47.0\ \mu\mathrm{M} \pm 8.8\ \mu\mathrm{M}$. This $k_{cat,f}$ is more than four times higher than the one reported by Zhao *et al.*[12] As an additional control, the activity of ALD after all NMR experiments and prolonged exposure to room temperature was measured at 10 mM FBP, which yielded an activity of at least $2.19\ \mathrm{s}^{-1} \pm 0.09\ \mathrm{s}^{-1}$ or higher. A similar measurement was performed for ALD after prolonged exposure to room temperature using 10 mM FBP in the presence of tracer molecule mixture (5 mg/mL dioxane and 2.5 mg/mL 18-crown-6), which leads to an activity of $2.01\ \mathrm{s}^{-1} \pm 0.08\ \mathrm{s}^{-1}$.

$$v = \frac{d[FBP]}{dt} = \frac{v_{max,f}\dfrac{[FBP]}{K_{FBP}} - v_{max,r}\dfrac{[G3P][DHAP]}{K_{G3P}K_{DHAP}}}{1 + \dfrac{[FBP]}{K_{FBP}} + \dfrac{[G3P]}{K_{G3P}} + \dfrac{[DHAP]}{K_{DHAP}} + \dfrac{[G3P][DHAP]}{K_{G3P}K_{DHAP}}} \tag{A.4}$$

$$v = \frac{v_{max,f}[FBP]}{K_{FBP}+[FBP]} \quad \text{for} \quad [G3P] = [DHAP] = 0 \tag{A.5}$$

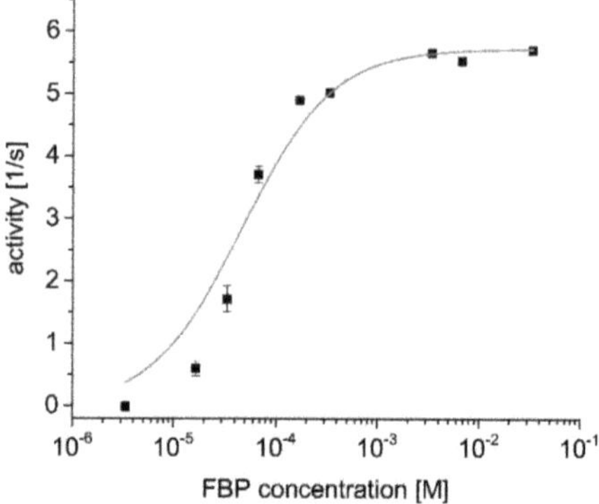

Figure A.2. ALD activity assay. Michaelis-Menten plot of ALD activity in Tris-d_{11} buffer. Each data point results from three measurements.

At the FBP concentrations of the diffusion PFG-NMR experiments the equilibrium is quickly reached (within a few minutes, assuming that $v_{max,r}$ is of similar magnitude than $v_{max,f}$). The forward reaction speed at equilibrium can be estimated by calculating the $v_{eq,f}/v_{max,r}$ as Zhang *et al.* have already shown:[106]

$$\frac{v_{eq,f}}{v_{max,f}} = \frac{\frac{[FBP]_{eq}}{K_{FBP}}}{1 + \frac{[FBP]_{eq}}{K_{FBP}} + \frac{[G3P]_{eq}}{K_{G3P}} + \frac{[DHAP]_{eq}}{K_{DHAP}} + \frac{[G3P]_{eq}[DHAP]_{eq}}{K_{G3P}K_{DHAP}}} \tag{A.6}$$

$[FBP]_{eq}$, $[G3P]_{eq}$ and $[DHAP]_{eq}$ can be calculated using the equilibrium constant reported in the literature $K_{eq} = 8.0 \cdot 10^{-5}$ M.[230] For an initial substrate concentration of $[FBP]_0 =$ 50 mM the equilibrium forward velocity can be calculated to be 77 % of $v_{max,f}$ using the Michaelis constants by Callens *et al.* ($K_{FBP} = 18$ µM, $K_{G3P} = 67$ µM, $K_{DHAP} = 15$ µM)[231] and 99 % using the Michaelis constants by Rutter *et al.* ($K_{FBP} = 61$ µM, $K_{G3P} = 1.1$ mM, $K_{DHAP} =$ 2.1 mM).[232] Therefore, we assume that even after the reaction reached equilibrium, sufficient forward reaction occurs to test the diffusive behavior of ALD during FBP conversion.

A.1.3 Detailed Settings for the PFG-NMR Experiments

For all PFG-NMR experiments a stimulated echo sequence has been used with the parameters listed in Tab. A.1. Special care was taken to utilize a very fast pulse sequence for the experiments where the ALD diffusion was monitored during FBP turnover. All the gradients used in the experiments of this paper are listed in Tab. A.2. Only high gradients have been selected for measurements of ALD + FBP, since these measurements were time-sensitive and to remove contribution from smaller molecules. For comparability, the gradient range of the measurements ALD + PP and ALD + FBP was selected to be the same for the nonlinear regression, namely 4.242-12.000 T/m.

Table A.1. Parameters for the PFG-NMR experiments presented in Fig. 8.3, 8.4 and 8.5.

	experiments		
parameter	**tracers**	**ALD + PP**	**ALD + FBP**
δ_G [ms]	1	0.6	0.6
Δ [ms]	10	8	8
length of FID [s]	1.049	0.262	0.262
recovery delay [s]	1.00	0.50	0.25
number of scans	16	128	2x64
number of dummy scans	1	1	1
number of gradients	16	16	8
total time	8 min 50 s	26 min 30 s	9 min

Table A.2. Gradients G used in the PFG-NMR experiments presented in Fig. 8.3, 8.4 and 8.5.

		experiments	
gradient	**tracers**	**ALD + PP**	**ALD + FBP**
number	***G* [T/m]**	***G* [T/m]**	***G* [T/m]**
1	0.100	3.000	4.242
2	0.250	4.242	5.999
3	0.400	5.195	7.347
4	0.550	5.999	8.484
5	0.700	6.707	9.485
6	0.850	7.347	10.390
7	1.000	7.936	11.220
8	1.150	8.484	12.000
9	1.300	8.998	
10	1.450	9.485	
11	1.600	9.948	
12	1.750	10.390	
13	1.900	10.810	
14	2.100	11.220	
15	2.250	11.620	
16	2.400	12.000	

A.1.4 Detailed Results of the PFG-NMR Experiments

For all PFG-NMR experiments the signals of the individual molecules were integrated for each gradient to calculate $I(G)$. The intervals and gradients used for those integrations are listed at the bottom of the individual tables. Subsequently, a nonlinear regressions of $I(G)$ was performed, which resulted in the values of I_0 and D. All the results for D are listed together with the standard deviation and the R^2 value obtained from the nonlinear regression. All regressions had an $R^2 > 0.97$.

Diffusion measurements of solutions in the presence of active aldolase

Fig. A.3 presents the data and regression of all tracer experiments performed. The differences in the signal attenuation for each molecule are seen in the figures. The results of the nonlinear regression are listed in Tab. A.3. Here all R^2 values are greater then 0.9995.

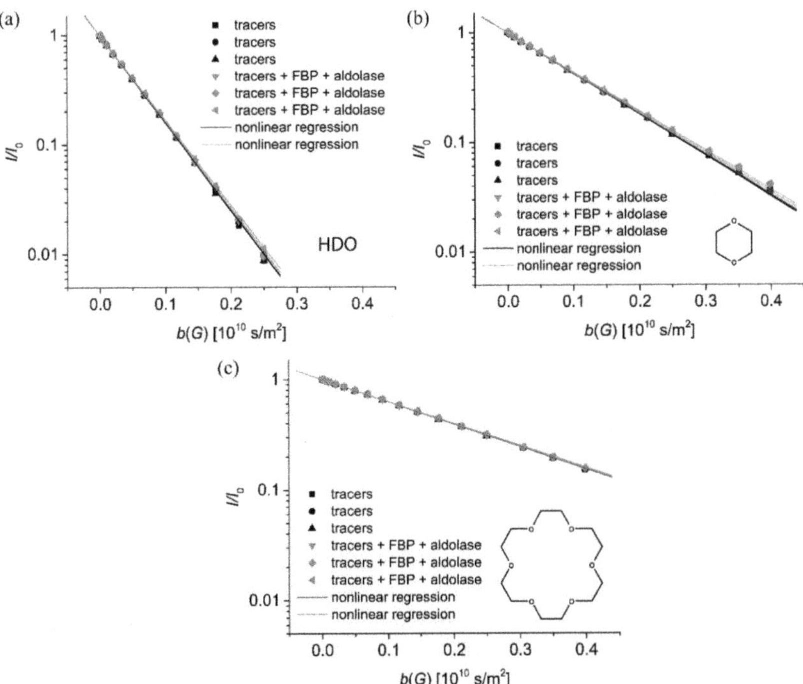

Figure A.3. Regressions of the tracer molecule signal attenuation during the experiments presented in Fig. 8.3. (a) HDO signal between δ=4.680 ppm and 4.750 ppm. (b) dioxane signal between δ=3.680 ppm and 3.705 ppm. (c) 18-crown-6 signal between δ=3.604 ppm and 3.635 ppm. 5 mg/mL dioxane and 2.5 mg/mL 18-crown-6 were used for the control experiments (black). Additionally, 10 mM FBP and 10 nM ALD were present during the active ALD experiments (red).

Table A.3. Diffusion coefficients of tracer molecules from PFG-NMR experiments performed in active ALD solutions.

c(FBP) [mM]	*c*(aldolase) [nM]	*D*(HDO)* [10^{-10} m^2/s]	R^2	*D*($C_4H_8O_2$)** [10^{-10} m^2/s]	R^2	*D*($C_{12}H_{24}O_6$)*** [10^{-10} m^2/s]	R^2
0	0	18.4 ± 0.1	0.9999	8.55 ± 0.05	0.9999	4.68 ± 0.02	0.9999
0	0	18.1 ± 0.2	0.9999	8.53 ± 0.05	0.9999	4.66 ± 0.03	0.9998
0	0	18.4 ± 0.2	0.9999	8.60 ± 0.05	0.9999	4.68 ± 0.03	0.9998
0	10	18.3 ± 0.1	1.0000	8.58 ± 0.05	0.9999	4.67 ± 0.02	0.9999
0	10	18.2 ± 0.1	1.0000	8.49 ± 0.04	0.9999	4.65 ± 0.03	0.9999
0	10	18.2 ± 0.1	0.9999	8.56 ± 0.05	0.9999	4.65 ± 0.03	0.9999
0.1	0	18.4 ± 0.2	0.9999	8.57 ± 0.06	0.9999	4.69 ± 0.03	0.9999
0.1	0	17.9 ± 0.1	1.0000	8.46 ± 0.05	0.9999	4.60 ± 0.02	1.0000
0.1	0	18.1 ± 0.2	0.9999	8.58 ± 0.04	0.9999	4.64 ± 0.03	0.9999
0.1	10	18.4 ± 0.1	0.9999	8.45 ± 0.07	0.9998	4.71 ± 0.02	0.9999
0.1	10	18.2 ± 0.1	0.9999	8.47 ± 0.05	0.9999	4.66 ± 0.01	1.0000
0.1	10	18.3 ± 0.1	1.0000	8.57 ± 0.07	0.9998	4.70 ± 0.02	1.0000
1	0	18.2 ± 0.1	0.9999	8.48 ± 0.05	0.9999	4.64 ± 0.02	0.9999
1	0	18.2 ± 0.2	0.9999	8.46 ± 0.06	0.9998	4.65 ± 0.04	0.9997
1	0	18.2 ± 0.1	0.9999	8.43 ± 0.05	0.9999	4.67 ± 0.03	0.9999
1	10	18.1 ± 0.2	0.9999	8.47 ± 0.05	0.9999	4.69 ± 0.02	0.9999
1	10	18.1 ± 0.1	0.9999	8.50 ± 0.06	0.9998	4.63 ± 0.02	0.9999
1	10	18.2 ± 0.1	0.9999	8.57 ± 0.05	0.9999	4.70 ± 0.02	0.9999
10	0	18.1 ± 0.2	0.9999	8.31 ± 0.06	0.9998	4.62 ± 0.04	0.9997
10	0	17.6 ± 0.1	1.0000	8.16 ± 0.05	0.9999	4.54 ± 0.01	1.0000
10	0	18.1 ± 0.1	0.9999	8.49 ± 0.07	0.9998	4.70 ± 0.03	0.9998
10	10	18.1 ± 0.2	0.9999	8.47 ± 0.09	0.9995	4.66 ± 0.03	0.9998
10	10	17.8 ± 0.2	0.9999	8.35 ± 0.05	0.9999	4.64 ± 0.03	0.9999
10	10	17.8 ± 0.2	0.9998	8.23 ± 0.07	0.9998	4.59 ± 0.05	0.9995

*gradients 1-13, δ=4.680-4.750 ppm

**gradients 1-16, δ=3.680-3.705 ppm

***gradients 1-16, δ=3.604-3.635 ppm

Diffusion of the enzyme aldolase during interaction with its inhibitor pyrophosphate

Diffusion coefficients are listed in Tab. A.4 for all experiments involving the interaction of ALD with its inhibitor pyrophosphate (PP).

Table A.4. Diffusion coefficients of ALD PFG-NMR experiments performed during interaction with its inhibitor PP.

c(PP) [mM]	*c*(aldolase) [µM]	*D*(aldolase)* [10^{-11} m^2/s]	R^2
0	11.9	4.92 ± 0.14	0.993
0	11.9	4.60 ± 0.18	0.987
0	11.9	4.62 ± 0.17	0.988
6	11.9	4.51 ± 0.14	0.992
6	11.9	4.51 ± 0.19	0.986
6	11.9	4.60 ± 0.11	0.995

*gradients 2-16, δ=0.5-1.6 ppm

Diffusion of the enzyme aldolase during substrate conversion

In Fig. A.4 the 1D-^{1}H-NMR spectrum of FBP, the substrate of ALD, is shown. FBP shows no proton NMR signal in the interval between δ=0.5 ppm and 1.6 ppm, and therefore does not interfere with the aliphatic ALD signal analyzed in this paper. It is also seen that the FBP spectrum in Fig. A.4 exhibits more peaks than would be expected for a simple furanose structure. The additional peaks arise because FBP is in equilibrium between its furanose and keto form. Interestingly, the keto form of FBP shows a higher reaction rate with ALD than its furanose form.[233]

The results of the nonlinear regression of all ALD diffusion experiments during substrate conversion are listed in Tab. A.5.

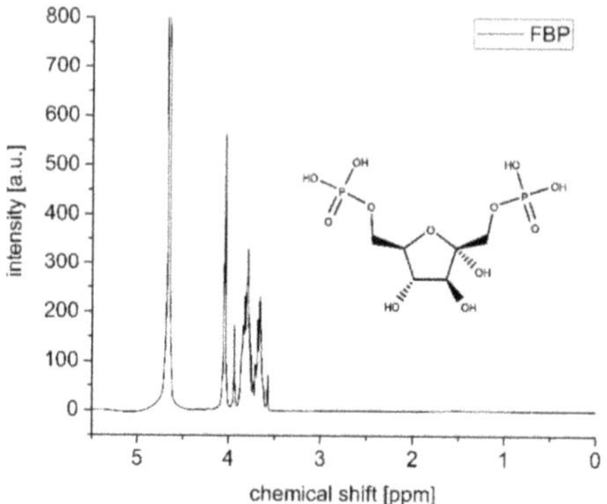

Figure A.4. ^{1}H-NMR spectrum of 50 mM FBP in Tris-d_{11} buffer and chemical structure of FBP.

Table A.5. Diffusion coefficients of ALD PFG-NMR experiments performed during conversion of its substrate FBP.

c(FBP) [mM]	*c*(aldolase) [µM]	*D*(aldolase)* [10^{-11} m²/s]	R^2
0	11.9	4.45 ± 0.22	0.992
0	11.9	4.81 ± 0.29	0.989
0	11.9	4.66 ± 0.25	0.991
50	11.9	4.39 ± 0.21	0.993
50	11.9	4.49 ± 0.23	0.987
50	11.9	4.81 ± 0.37	0.986

*gradients 1-8, δ=0.5-1.6 ppm

A.2 Enzyme-Phage-Colloid Supporting Information

This appendix is reprinted with minor formatting adjustments with permission of ACS publications from the Supporting Information of Ref. 3 (Chapter 9).

A.2.1. Genetic Modification of M13 Bacteriophages Templates

Modification of the pIII minor coat protein

The pIII coat protein was modified by the insertion of a synthetic DNA fragment using the restriction sites *Knp*I and *Eag*I (Fig. A.5). M13KE DNA was digested with *Kpn*I-HF and *Eag*I-HF (both NEB) and the DNA was purified by gel extraction. The DNA for insertion was synthesized as complementary oligonucleotides (Biomers, Germany) with internal *Knp*I and *Eag*I restriction sites (underlined):

p3HT_for 5′ A GTGGTACCTT TCTATTCTCA CTCTCATCAT CACCATCACC ACCTGGTTCC GCGTGGATCC TCGGCCGAAA 3′ and p3HT_rev 5′ T TTCGGCCGA GGATCCAGCG GGAACCAGGT GGTGATGGTG ATGATGAGAG TGAGAATAGA AAGGTACCACT 3′. The oligonucleotides were hybridized (5 min., 95 °C; to 25 °C with 2 °C/sec cooling rate) and digested with *Knp*I and *Eag*I. The reaction was purified with MinElute reaction cleanup kit (Qiagen). The insert was ligated into the linearized vector DNA at 20°C for one hour using T4 DNA ligase (NEB, USA). The ligation product was transferred into *E. coli* ER2738 cells by chemical transformation. Plaque forming units (pfu) were screened on LB agar plates containing IPTG and X-Gal and analyzed by DNA sequencing.

Modification of the pVII minor coat proteins

The pVII and pIX minor coat protein genes were modified by an adapted version of the inverse polymerase chain reaction (iPCR) method.[234] Abutting primers were designed harboring the his-tag and the protease restriction site as well as an addition *Nhe*I restriction site. For the modification of the pVII gene the primer pair p7HXa_plus 5′AATTGTGCTA GCGTGGTGAT GGTGATGATG CATGTTACTT AGCCGGAACG AGGCGCAGAC3′ and p7HXa_minus 5′TCTGCGCCGC TAGCATTGAT GGACGTATGG AGCAGGTCGC GGATTTCGAC ACAATTTATC 3′ were used. M13KE DNA was applied as template for the amplification. Remaining nucleotides and enzymes were removed from the obtained PCR product with the GeneJET PCR purification kit (Thermo Scientific). The purified PCR product was *NheI* digested, gel purified and re-circularized by ligation at 20 °C for one hour. The ligated DNA was transfected into competent *E. coli* ER2738 cells as described for the pIII gene modification.

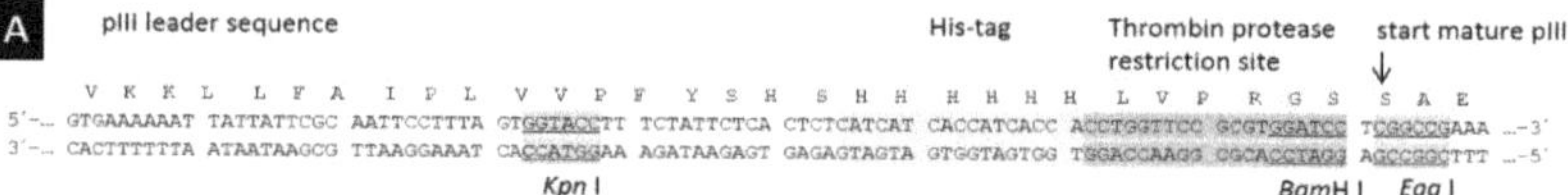

Figure A.5. Cloning strategy for phage templates. DNA sequences of genetically modified M13 and fd phage for the expression of a His-tag and protease restriction site. A) M13 p3 His-tag + Thrombin protease restriction site (M13 3HT).

A.2.2 Scanning and Transmission Electron Micrographs of Phage-Colloid Constructs and References

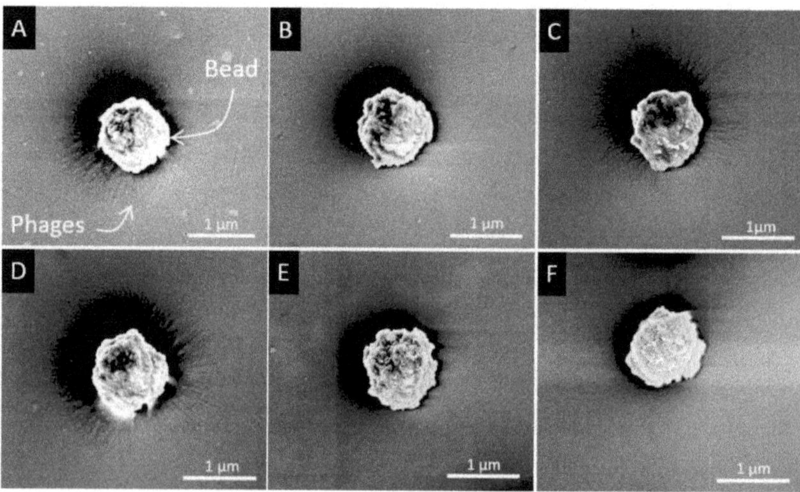

Figure A.6. Coupling of M13 and fd phages to magnetic particles via the genetically engineered His-tag. SEM micrographs of A) M13 3HT, B) M13 wild type (wt), C) fd 3 HT, D) fd 7 HX, E) fd Y21M and F) magnetic particle without phage templates. Only phages expressing a His-tag (A, C, D) showed binding to the magnetic particle, which is visible by the corona surrounding the magnetic particle. Non-modified phages (B, E) did not show interaction with the particle.

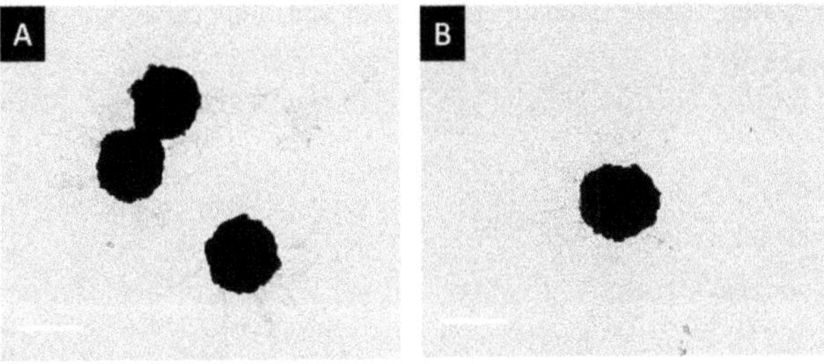

Figure A.7. Transmission electron micrograph of the magnetic carrier particles functionalized with M13 3HT bacteriophages and urease. Scale bar 1 µm. Images taken with the support of Dr. Philipp Erdmann and Chuan Liu at the MPI of Biochemistry.

A.2.3 Determination of Enzyme Loading on Phage-Colloids

For all experiments, urease type IX from *Canavalia ensiformis* (jack bean) from Sigma Aldrich was used. The correction factor is determined to correlate the weight of solid to the amount of pure urease in the product. The urease content is measured by UV-Vis spectroscopy at 280 nm with the molar extinction coefficient (ε) of the urease homo hexamer (ε= 309,100 $M^{-1}*cm^{-1}$) calculated[228] from the PDB entry 3LA4 sequence.[117] Four urease standard solutions with a concentration of 10 mg solid*mL^{-1} in (ddH_2O) are prepared. 1 g of powder contains 120 ± 10 mg of urease.

The powder is a mixture of urease and precanavalin.[115] In polyacrylamide gel electrophoresis (PAGE), the urease subunit (~100 kDa) and the precanavalin (~50 kDa) are separated bands (Fig. A.8 a). The PAGE is further analyzed with ImageJ. The band intensities were plotted and the fraction of the urease and precanavalin band intensity is calculated to ~ 0.65. Together with the urease content of the solid, the correction factor 0.077 ± 0.007 is calculated converting the µg solid to µg urease. The band intensities are plotted over the urease amount (Fig. A.8 b). The values are fitted with a linear model and the amount of urease in each samples is calculated. Urease standard samples and calibration curves are performed with each individual PAGE.

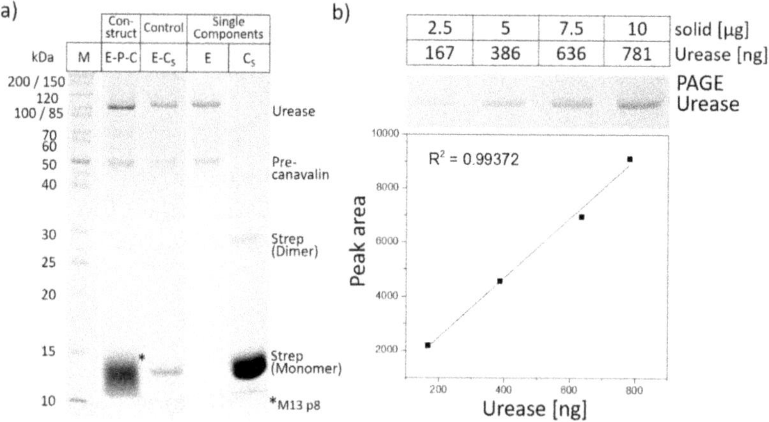

Figure A.8. Determination of urease content. a) Polyacrylamide gel electrophoresis (PAGE) of E-P-C construct, colloids with directly coupled urease as control (E-C_S), and the single components: urease enzyme (E) and streptavidin-coated colloids (C_S) for direct urease coupling. b) Urease standard samples with 2.5, 5.0, 7.5, and 10 mg solid / lane and the corresponding absolute urease amount / lane. Samples are separated on 12.5 % gels by SDS-polyacrylamide gel electrophoresis (PAGE). The band intensities are evaluated with ImageJ. The contrast and brightness of PAGE pictures have been adjusted for better visibility.

The number of phages on the carrier particle was determined by UV-Vis spectroscopy. Therefore, the number of phages in the supernatant after the phage-carrier binding was determined and subtracted from the initial amount of phages in the experiment. Based on the measurements 77.1 ± 6.5 % of the genetically modified phages bind to the carrier particles. The concentration of the carrier particle stock solution was counted in a Neubauer counting chamber using 10^{-3} and 10^{-4} dilutions. The concentration was $4.42\times10^{10} \pm 9.3\times10^{9}$ beads per mL.

Table A.6. Characterization of enzyme-phage-carriers (E-P-C). The numbers are based on seven independent experiments each with 16 repeats in total. The amount of phages per bead was determined by UV-Vis spectroscopy and the concentration of urease by PAGE. The mean value (mean) and standard deviation (STAD) are given at the end of the table.

Sample	Phagen / 50 µl bead	Phagen / bead	% Phage binding	Urease/50 µL bead [µg]	Urease/50 µL bead [number]	Urease / bead
1_01	7.52E+11	340.4	75.2	0.92	1.01E+12	457.8
1_02	7.52E+11	340.4	75.2	1.41	1.56E+12	704.5
1_03	7.52E+11	340.4	75.2	1.38	1.52E+12	688.0
2_01	7.28E+11	329.6	72.8	1.29	1.43E+12	647.9
2_02	7.28E+11	329.6	72.8	1.22	1.35E+12	611.9
3_01	7.29E+11	329.8	72.9	1.81	2.00E+12	905.8
3_02	7.29E+11	329.8	72.9	1.79	1.98E+12	895.6
4_01	7.15E+11	323.3	71.5	1.20	1.33E+12	601.9
4_02	7.15E+11	323.3	71.5	1.23	1.36E+12	616.9
5_01	7.11E+11	321.6	71.1	1.06	1.17E+12	529.7
5_02	7.11E+11	321.6	71.1	1.51	1.67E+12	757.4
6_01	8.63E+11	390.7	86.3	0.95	1.05E+12	476.4
6_02	8.63E+11	390.7	86.3	2.29	2.53E+12	1145.4
6_03	8.63E+11	390.7	86.3	1.95	2.15E+12	973.5
7_01	8.61E+11	389.6	86.1	1.45	1.61E+12	727.3
7_02	8.61E+11	389.6	86.1	1.73	1.91E+12	865.0
Mean	7.71E+11	348.8	77.1	1.4	1.60E+12	725.3
STAD	6.50E+10	29.5	6.5	0.4	4.19E+11	189.7

A.2.4 Enzyme Activity Assay

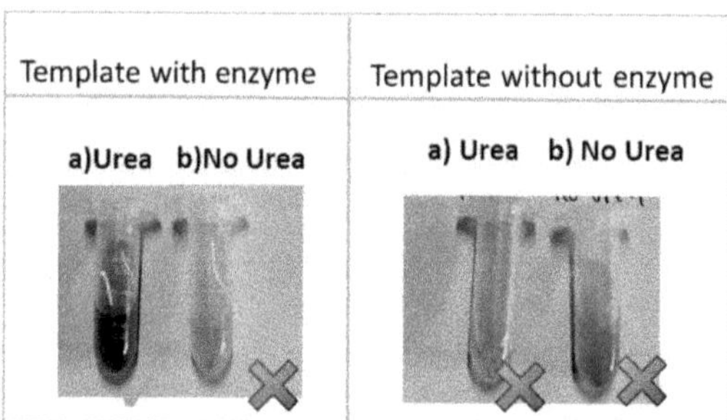

Figure A.9. Urease activity of the enzyme-phage-colloid (E-P-C) is determined by a colorimetric assay. Functional urease enzyme catalyzes the hydrolysis of urea into carbon dioxide and ammonia. The ammonia induces a color change from yellow to light blue/green indicating an active enzyme. Only samples containing urease immobilized on the phage templates and urea showed catalytic activity (template with enzyme, a).

Samples missing either urea or the enzyme did not show catalytic activity, indicating no unspecific degradation of urea by the phage-colloid construct. Note that the cloudy appearance of the reaction solution is due to the carrier particles.

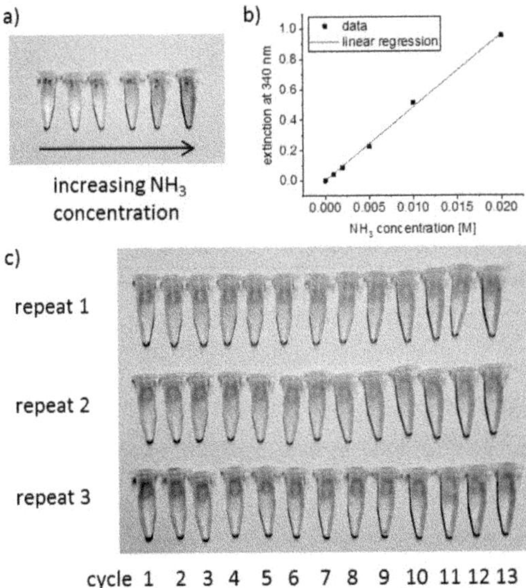

Figure A.10. Calibration curve and colorimetric assay for the reusability of the E-P-C. a) Calibration experiments with varying ammonium concentrations. b) Calibration curve and linear regression. c) Stability and reusability experiment with three repeats and 13 cycles each. The total experimental time is >8 h. In all samples, the enzyme was catalytically active after each transfer and over the total time of the experiment.

A.2.5 Enzyme Kinetics Calculations

Using the *in situ* coupled enzymatic assay described in the materials and methods section (Chapter 9.5), the activity of free urease and urease on E-P-Cs has been determined. Together with the urease concentrations determined by UV spectroscopy and PAGE, we were able to calculate the turnover rate k_{cat} for both samples.

Table A.7. Coupled activity assay for urease free in solution and coupled to the E-P-Cs. Activity and k_{cat} values are given per consumed urea molecule.

sample	activity (/µM*s^{-1})	urease conc. (/nM)	k_{cat} (/s^{-1})
free urease	0.161± 0.004	0.741 ± 0.002	218 ± 5
urease on E-P-Cs	0.0768 ± 0.0094	0.175 ± 0.054	438 ± 145

A2.6 Micropump System

To ensure that the fluid flow is indeed caused by the catalytic activity of the E-P-C, the following control experiments are conducted. The flow velocities are an average value over three or more independent measurements at different locations of the active site or different chambers.

First, enzymes were directly bound onto the surface of a carrier particle. Therefore, biotin labeled urease (EZ-Link NHS-Biotin, Thermo Scientific) is immobilized on streptavidin decorated magnetic particles (Dynabeads MyOne, Novex) resulting in the enzyme-colloid (E-C) systems. In order to set up reference experiments with the same concentration of the E-C and the E-P-C system, the extinction coefficients of both colloids (Dynabead His-tag® and MyOne®) are determined and solutions with the same colloid concentration are prepared. The pumping speed of the E-C system is determined to be 3.04 ± 0.4 µm*s^{-1} on the bottom and 3.02 ± 0.5 µm*s^{-1} on the top from three independent experiments (Fig. A.11).

Free urease is added to the fluid chamber with the non-functionalized phage-colloid (P-C without urease) at the bottom. Under the same reaction conditions, no flow of the tracer particles is observed (Fig. A.11). Therefore, we can exclude that the flow is caused by the catalytic activity of the enzyme alone, but is linked to the localization of the enzymatic activity.

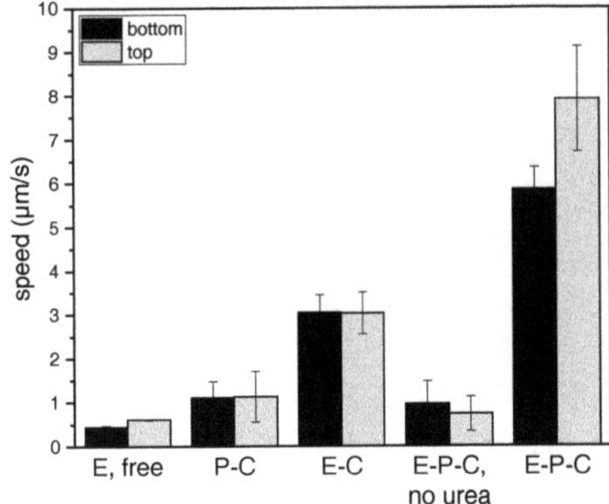

Figure A.11. Speed velocities of various micropump systems. Micropump systems were set up with (i) enzyme (urease) free in solution (E, free), (ii) phage-colloids without enzyme (P-C), (iii) biotin-labelled enzymes bound to streptavidin-coated colloids (E-C), (iv) the enzyme-phage-colloid (E-P-C) in the absence of substrate (E-P-C, no urea) and (v) the E-P-C with substrate.

References

1. Günther, J.-P., Börsch, M. & Fischer, P. Diffusion Measurements of Swimming Enzymes with Fluorescence Correlation Spectroscopy. *Acc. Chem. Res.* **51**, 1911–1920 (2018).
2. Günther, J.-P., Majer, G. & Fischer, P. Absolute Diffusion Measurements of Active Enzyme Solutions by NMR. *J. Chem. Phys.* **150**, 124201 (2019).
3. Alarcón-Correa, M. *et al.* Self-Assembled Phage-Based Colloids for High Localized Enzymatic Activity. *ACS Nano* **13**, 5810–5815 (2019).
4. Günther, J.-P. *et al.* Comment on 'Boosted Molecular Mobility During Common Chemical Reactions'. (2020) doi:10.26434/chemrxiv.13023164.v1.
5. Günther, J.-P. *et al.* Comment on "Boosted molecular mobility during common chemical reactions". *Science* **371**, eabe8322 (2021).
6. Thomas, J., Miller, M., McClendon, Z. & Schyman, G. *BioShock 2*. (2K Marin, 2010).
7. Muddana, H. S., Sengupta, S., Mallouk, T. E., Sen, A. & Butler, P. J. Substrate Catalysis Enhances Single-Enzyme Diffusion. *J. Am. Chem. Soc.* **132**, 2110–2111 (2010).
8. Riedel, C. *et al.* The heat released during catalytic turnover enhances the diffusion of an enzyme. *Nature* **517**, 227–230 (2015).
9. Illien, P. *et al.* Exothermicity Is Not a Necessary Condition for Enhanced Diffusion of Enzymes. *Nano Lett.* **17**, 4415–4420 (2017).
10. Jee, A.-Y., Dutta, S., Cho, Y.-K., Tlusty, T. & Granick, S. Enzyme leaps fuel antichemotaxis. *Proc. Natl. Acad. Sci. U.S.A.* **115**, 14–18 (2018).
11. Börsch, M. *et al.* Conformational changes of the H+-ATPase from Escherichia coli upon nucleotide binding detected by single molecule fluorescence. *FEBS Lett.* **437**, 251–254 (1998).
12. Zhao, X. *et al.* Substrate-driven chemotactic assembly in an enzyme cascade. *Nat. Chem.* **10**, 311–317 (2018).
13. Wang, H. *et al.* Boosted Molecular Mobility During Common Chemical Reactions. *Science* **369**, 537–541 (2020).
14. Feng, M. & Gilson, M. K. A Thermodynamic Limit on the Role of Self-Propulsion in Enhanced Enzyme Diffusion. *Biophys. J.* **116**, 1898–1906 (2019).
15. Purcell, E. M. Life at low Reynolds number. *Am. J. Phys.* **45**, 3–11 (1977).
16. Happel, J. & Brenner, H. *Low Reynolds number hydrodynamics*. vol. 1 (Springer Netherlands, 1981).
17. Turner, L., Ryu, W. S. & Berg, H. C. Real-Time Imaging of Fluorescent Flagellar Filaments. *J. Bacteriol.* **182**, 2793–2801 (2000).
18. Brennen, C. & Winet, H. Fluid Mechanics of Propulsion by Cilia and Flagella. *Annu. Rev. Fluid Mech.* **9**, 339–398 (1977).
19. Kaupp, U. B. & Alvarez, L. Sperm as microswimmers – navigation and sensing at the physical limit. *Eur. Phys. J. Spec. Top.* **225**, 2119–2139 (2016).

20. Layton, D. The original observations of Brownian motion. *J. Chem. Educ.* **42**, 367 (1965).

21. Huang, R. *et al.* Direct observation of the full transition from ballistic to diffusive Brownian motion in a liquid. *Nat. Phys.* **7**, 576–580 (2011).

22. Price, W. S. *NMR Studies of Translational Motion*. (Cambridge University Press, 2009).

23. Sengupta, S. *et al.* Enzyme Molecules as Nanomotors. *J. Am. Chem. Soc.* **135**, 1406–1414 (2013).

24. Sengupta, S. *et al.* DNA Polymerase as a Molecular Motor and Pump. *ACS Nano* **8**, 2410–2418 (2014).

25. Wu, F., Pelster, L. N. & Minteer, S. D. Krebs cycle metabolon formation: metabolite concentration gradient enhanced compartmentation of sequential enzymes. *Chem. Commun.* **51**, 1244–1247 (2015).

26. Einstein, A. Über die von der molekularkinetischen Theorie der Wärme geforderte Bewegung von in ruhenden Flüssigkeiten suspendierten Teilchen. *Ann. Phys.* **322**, 549–560 (1905).

27. Sutherland, W. LXXV. A dynamical theory of diffusion for non-electrolytes and the molecular mass of albumin. *Phil. Mag.* **9**, 781–785 (1905).

28. Bai, X. & Wolynes, P. G. On the hydrodynamics of swimming enzymes. *J. Chem. Phys.* **143**, 165101 (2015).

29. Moran, J. L. & Posner, J. D. Phoretic Self-Propulsion. *Annu. Rev. Fluid Mech.* **49**, 511–540 (2017).

30. Kapral, R. Perspective: Nanomotors without moving parts that propel themselves in solution. *J. Chem. Phys.* **138**, 020901 (2013).

31. Colberg, P. H., Reigh, S. Y., Robertson, B. & Kapral, R. Chemistry in Motion: Tiny Synthetic Motors. *Acc. Chem. Res.* **47**, 3504–3511 (2014).

32. Paxton, W. F. *et al.* Catalytic Nanomotors: Autonomous Movement of Striped Nanorods. *J. Am. Chem. Soc.* **126**, 13424–13431 (2004).

33. Ebbens, S., Tu, M.-H., Howse, J. R. & Golestanian, R. Size dependence of the propulsion velocity for catalytic Janus-sphere swimmers. *Phys. Rev. E* **85**, 020401 (2012).

34. Ma, X. & Sánchez, S. Bio-catalytic mesoporous Janus nano-motors powered by catalase enzyme. *Tetrahedron* **73**, 4883–4886 (2017).

35. Lee, T.-C. *et al.* Self-Propelling Nanomotors in the Presence of Strong Brownian Forces. *Nano Lett.* **14**, 2407–2412 (2014).

36. Santiago, I., Jiang, L., Foord, J. & Turberfield, A. J. Self-propulsion of catalytic nanomotors synthesised by seeded growth of asymmetric platinum–gold nanoparticles. *Chem. Commun.* **54**, 1901–1904 (2018).

37. Howse, J. R. *et al.* Self-Motile Colloidal Particles: From Directed Propulsion to Random Walk. *Phys. Rev. Lett.* **99**, 048102 (2007).

38. Brown, A. T., Poon, W. C. K., Holm, C. & Graaf, J. de. Ionic screening and dissociation are crucial for understanding chemical self-propulsion in polar solvents. *Soft Matter* **13**, 1200–1222 (2017).

39. Patiño, T., Arqué, X., Mestre, R., Palacios, L. & Sánchez, S. Fundamental Aspects of Enzyme-Powered Micro- and Nanoswimmers. *Acc. Chem. Res.* **51**, 2662–2671 (2018).

40. Schattling, P., Thingholm, B. & Städler, B. Enhanced Diffusion of Glucose-Fueled Janus Particles. *Chem. Mater.* **27**, 7412–7418 (2015).

41. Schattling, P. S., Ramos-Docampo, M. A., Salgueiriño, V. & Städler, B. Double-Fueled Janus Swimmers with Magnetotactic Behavior. *ACS Nano* **11**, 3973–3983 (2017).

42. Ma, X. *et al.* Enzyme-Powered Hollow Mesoporous Janus Nanomotors. *Nano Lett.* **15**, 7043–7050 (2015).

43. Ma, X., Wang, X., Hahn, K. & Sánchez, S. Motion Control of Urea-Powered Biocompatible Hollow Microcapsules. *ACS Nano* **10**, 3597–3605 (2016).

44. Tang, S. *et al.* Enzyme-powered Janus platelet cell robots for active and targeted drug delivery. *Sci. Robot.* **5**, eaba6137 (2020).

45. Hortelão, A. C., Patiño, T., Perez-Jiménez, A., Blanco, À. & Sánchez, S. Enzyme-Powered Nanobots Enhance Anticancer Drug Delivery. *Adv. Funct. Mater.* **28**, 1705086 (2018).

46. Echeverria, C., Togashi, Y., Mikhailov, A. S. & Kapral, R. A mesoscopic model for protein enzymatic dynamics in solution. *Phys. Chem. Chem. Phys.* **13**, 10527–10537 (2011).

47. Kuser, P., Cupri, F., Bleicher, L. & Polikarpov, I. Crystal structure of yeast hexokinase PI in complex with glucose: A classical "induced fit" example revised. *Proteins* **72**, 731–740.

48. Kuser, P. R., Krauchenco, S., Antunes, O. A. C. & Polikarpov, I. The High Resolution Crystal Structure of Yeast Hexokinase PII with the Correct Primary Sequence Provides New Insights into Its Mechanism of Action. *J. Biol. Chem.* **275**, 20814–20821 (2000).

49. Vale, R. D. The Way Things Move: Looking Under the Hood of Molecular Motor Proteins. *Science* **288**, 88–95 (2000).

50. Sielaff, H., Yanagisawa, S., Frasch, W. D., Junge, W. & Börsch, M. Structural Asymmetry and Kinetic Limping of Single Rotary F-ATP Synthases. *Molecules* **24**, 504 (2019).

51. Jiang, L., Santiago, I. & Foord, J. Observation of nanoimpact events of catalase on diamond ultramicroelectrodes by direct electron transfer. *Chem. Commun.* **53**, 8332–8335 (2017).

52. Jee, A.-Y., Cho, Y.-K., Granick, S. & Tlusty, T. Catalytic enzymes are active matter. *Proc. Natl. Acad. Sci. U.S.A.* **115**, E10812–E10821 (2018).

53. Goldberg, R. N. & Tewari, Y. B. Thermodynamics of Enzyme-Catalyzed Reactions. Part 3. Hydrolases. *J. Phys. Chem. Ref. Data* **23**, 1035–1103 (1994).

54. Ramaiya, A., Roy, B., Bugiel, M. & Schäffer, E. Kinesin rotates unidirectionally and generates torque while walking on microtubules. *Proc. Natl. Acad. Sci. U.S.A.* **114**, 10894–10899 (2017).

55. Golestanian, R. Enhanced Diffusion of Enzymes that Catalyze Exothermic Reactions. *Phys. Rev. Lett.* **115**, 108102 (2015).

56. Tsekouras, K. *et al.* Comment on 'Enhanced Diffusion of Enzymes that Catalyze Exothermic Reactions' by R.Golestanian. *arXiv:1608.05433 [physics]* (2016).

57. Golestanian, R. Reply to Comment on 'Enhanced diffusion of enzymes that catalyze exothermic reactions'. *arXiv:1608.07469 [physics, q-bio]* (2016).

58. Illien, P., Adeleke-Larodo, T. & Golestanian, R. Diffusion of an enzyme: The role of fluctuation-induced hydrodynamic coupling. *EPL* **119**, 40002 (2017).

59. Kondrat, S. & Popescu, M. N. Brownian dynamics assessment of enhanced diffusion exhibited by 'fluctuating-dumbbell enzymes'. *Phys. Chem. Chem. Phys.* **21**, 18811–18815 (2019).

60. Whitford, P. C., Gosavi, S. & Onuchic, J. N. Conformational Transitions in Adenylate Kinase: Allosteric Communication Reduces Misligation. *J. Biol. Chem.* **283**, 2042–2048 (2008).

61. Koumura, N., Zijlstra, R. W. J., van Delden, R. A., Harada, N. & Feringa, B. L. Light-driven monodirectional molecular rotor. *Nature* **401**, 152–155 (1999).

62. García-López, V. *et al.* Unimolecular Submersible Nanomachines. Synthesis, Actuation, and Monitoring. *Nano Lett.* **15**, 8229–8239 (2015).

63. Pavlick, R. A., Dey, K. K., Sirjoosingh, A., Benesi, A. & Sen, A. A catalytically driven organometallic molecular motor. *Nanoscale* **5**, 1301–1304 (2013).

64. Dey, K. K. *et al.* Dynamic Coupling at the Ångström Scale. *Angew. Chem. Int. Ed.* **55**, 1113–1117 (2016).

65. MacDonald, T. S. C., Price, W. S., Astumian, R. D. & Beves, J. E. Enhanced Diffusion of Molecular Catalysts is Due to Convection. *Angew. Chem. Int. Ed.* **58**, 18864–18867 (2019).

66. Gessler, S., Randl, S. & Blechert, S. Synthesis and metathesis reactions of a phosphine-free dihydroimidazole carbene ruthenium complex. *Tetrahedron Lett.* **41**, 9973–9976 (2000).

67. Zhao, X. *et al.* Enhanced Diffusion of Passive Tracers in Active Enzyme Solutions. *Nano Lett.* **17**, 4807–4812 (2017).

68. Tornøe, C. W., Christensen, C. & Meldal, M. Peptidotriazoles on Solid Phase: [1,2,3]-Triazoles by Regiospecific Copper(I)-Catalyzed 1,3-Dipolar Cycloadditions of Terminal Alkynes to Azides. *J. Org. Chem.* **67**, 3057–3064 (2002).

69. Cussler, E. L. *Diffusion: Mass Transfer in Fluid Systems*. (Cambridge University Press, 2009).

70. Schwille, P., Kummer, S., Heikal, A. A., Moerner, W. E. & Webb, W. W. Fluorescence correlation spectroscopy reveals fast optical excitation-driven intramolecular dynamics of yellow fluorescent proteins. *Proc. Natl. Acad. Sci. U.S.A.* **97**, 151–156 (2000).

71. Noji, H., Yasuda, R., Yoshida, M. & Jr, K. K. Direct observation of the rotation of F1-ATPase. *Nature* **386**, 299–302 (1997).

72. Ernst, S., Düser, M. G., Zarrabi, N., Dunn, S. D. & Börsch, M. Elastic deformations of the rotary double motor of single FoF1-ATP synthases detected in real time by Förster resonance energy transfer. *Biochim. Biophys. Acta* **1817**, 1722–1731 (2012).

73. Lakowicz, J. R. *Principles of Fluorescence Spectroscopy*. (Springer, 2010).

74. Rigler, R. & Elson, E. S. *Fluorescence Correlation Spectroscopy: Theory and Applications*. (Springer Berlin Heidelberg, 2001).

75. Kettling, U., Koltermann, A., Schwille, P. & Eigen, M. Real-time enzyme kinetics monitored by dual-color fluorescence cross-correlation spectroscopy. *Proc. Natl. Acad. Sci. U.S.A.* **95**, 1416–1420 (1998).

76. Phillies, G. D. J. Fluorescence correlation spectroscopy and nonideal solutions. *Biopolymers* **14**, 499–508 (1975).

77. Schwille, P. & Haustein, E. Fluorescence Correlation Spectroscopy. An Introduction to its Concepts and Applications. *BTOL* (2002).

78. Günther, H. *NMR Spectroscopy: Basic Principles, Concepts and Applications in Chemistry*. (Wiley-VCH, 2013).

79. Bloembergen, N., Purcell, E. M. & Pound, R. V. Relaxation Effects in Nuclear Magnetic Resonance Absorption. *Phys. Rev.* **73**, 679–712 (1948).

80. Majer, G. *Die Methoden der Kernspinresonanz zum Studium der Diffusion von Wasserstoff in Metallen und intermetallischen Verbindungen*. (Cuvillier, 2000).

81. Johnson, C. S. Diffusion ordered nuclear magnetic resonance spectroscopy: principles and applications. *Prog. Nucl. Magn. Reson. Spectrosc.* **34**, 203–256 (1999).

82. Majer, G. & Zick, K. Accurate and absolute diffusion measurements of Rhodamine 6G in low-concentration aqueous solutions by the PGSE-WATERGATE sequence. *J. Chem. Phys.* **142**, 164202 (2015).

83. Tanner, J. E. Use of the Stimulated Echo in NMR Diffusion Studies. *J. Chem. Phys.* **52**, 2523–2526 (1970).

84. Stejskal, E. O. & Tanner, J. E. Spin Diffusion Measurements: Spin Echoes in the Presence of a Time-Dependent Field Gradient. *J. Chem. Phys.* **42**, 288–292 (1965).

85. Jerschow, A. & Müller, N. Suppression of Convection Artifacts in Stimulated-Echo Diffusion Experiments. Double-Stimulated-Echo Experiments. *J. Magn. Reson.* **125**, 372–375 (1997).

86. MacDonald, T. S. C., Price, W. S. & Beves, J. E. Time-Resolved Diffusion NMR Measurements for Transient Processes. *ChemPhysChem* **20**, 926–930 (2019).

87. Stetefeld, J., McKenna, S. A. & Patel, T. R. Dynamic light scattering: a practical guide and applications in biomedical sciences. *Biophys. Rev.* **8**, 409–427 (2016).

88. Qian, H., Sheetz, M. P. & Elson, E. L. Single particle tracking. Analysis of diffusion and flow in two-dimensional systems. *Biophys. J.* **60**, 910–921 (1991).

89. Xu, M., Ross, J. L., Valdez, L. & Sen, A. Direct Single Molecule Imaging of Enhanced Enzyme Diffusion. *Phys. Rev. Lett.* **123**, 128101 (2019).

90. Neuman, K. C. & Nagy, A. Single-molecule force spectroscopy: optical tweezers, magnetic tweezers and atomic force microscopy. *Nat Methods* **5**, 491–505 (2008).

91. Svoboda, K., Schmidt, C. F., Schnapp, B. J. & Block, S. M. Direct observation of kinesin stepping by optical trapping interferometry. *Nature* **365**, 721–727 (1993).

92. Cohen, A. E. & Moerner, W. E. Method for trapping and manipulating nanoscale objects in solution. *Appl. Phys. Lett.* **86**, 093109 (2005).

93. Fields, A. P. & Cohen, A. E. Electrokinetic trapping at the one nanometer limit. *Proc. Natl. Acad. Sci. U.S.A.* **108**, 8937–8942 (2011).

94. Guha, R. *et al.* Chemotaxis of Molecular Dyes in Polymer Gradients in Solution. *J. Am. Chem. Soc.* **139**, 15588–15591 (2017).

95. Ortega, A., Amorós, D. & García de la Torre, J. Prediction of Hydrodynamic and Other Solution Properties of Rigid Proteins from Atomic- and Residue-Level Models. *Biophys. J.* **101**, 892–898 (2011).

96. García de la Torre, J., Huertas, M. L. & Carrasco, B. Calculation of Hydrodynamic Properties of Globular Proteins from Their Atomic-Level Structure. *Biophys. J.* **78**, 719–730 (2000).

97. García de la Torre, J., Huertas, M. L. & Carrasco, B. HYDRONMR: Prediction of NMR Relaxation of Globular Proteins from Atomic-Level Structures and Hydrodynamic Calculations. *J. Magn. Reson.* **147**, 138–146 (2000).

98. De Duve, C. *Life evolving: molecules, mind, and meaning*. (Oxford University Press, 2002).

99. Pettersen, E. F. *et al.* UCSF Chimera—A visualization system for exploratory research and analysis. *J. Comput. Chem.* **25**, 1605–1612 (2004).

100. Rader, B. A. Alkaline Phosphatase, an Unconventional Immune Protein. *Front. Immunol.* **8**, 897 (2017).

101. McComb, R. B., Bowers, Jr., G. N. & Posen, S. *Alkaline Phosphatase*. (Springer, 2013).

102. Ghosh, K. *et al.* Crystal structure of rat intestinal alkaline phosphatase – Role of crown domain in mammalian alkaline phosphatases. *J. Struct. Biol.* **184**, 182–192 (2013).

103. Butterworth, P. J. Alkaline phosphatase. Biochemistry of mammalian alkaline phosphatases. *Cell Biochem. Funct.* **1**, 66–70 (1983).

104. Baykov, A. A., Evtushenko, O. A. & Avaeva, S. M. A malachite green procedure for orthophosphate determination and its use in alkaline phosphatase-based enzyme immunoassay. *Anal. Biochem.* **171**, 266–270 (1988).

105. Gefflaut, T. Class I Aldolases: Substrate Specificity, Mechanism, Inhibitors and Structural Aspects. *Prog. Biophys. molec. Biol.* **63**, 301–340 (1995).

106. Zhang, Y., Armstrong, M. J., Bassir Kazeruni, N. M. & Hess, H. Aldolase Does Not Show Enhanced Diffusion in Dynamic Light Scattering Experiments. *Nano Lett.* **18**, 8025–8029 (2018).

107. Jee, A.-Y., Chen, K., Tlusty, T., Zhao, J. & Granick, S. Enhanced Diffusion and Oligomeric Enzyme Dissociation. *J. Am. Chem. Soc.* **141**, 20062–20068 (2019).

108. Tolan, D. R., Schuler, B., Beernink, P. T. & Jaenicke, R. Thermodynamic Analysis of the Dissociation of the Aldolase Tetramer Substituted at One or Both of the Subunit Interfaces. *Biol. Chem.* **384**, (2003).

109. Woodfin, B. M. Substrate-induced dissociation of rabbit muscle aldolase into active subunits. *Biochem. Biophys. Res. Commun.* **29**, 288–293 (1967).

110. Beernink, P. T. & Tolan, D. R. Disruption of the aldolase A tetramer into catalytically active monomers. *Proc. Natl. Acad. Sci. U.S.A.* **93**, 5374–5379 (1996).

111. Kasprzak, A. A. & Kochman, M. Interaction of Fructose- 1,6-Bisphosphate Aldolase with Adenine Nucleotides. Binding of 5'-Mononucleotides and Phosphates to Rabbit Muscle Aldolase. *Eur. J. Biochem.* **104**, 443–450 (1980).

112. Blostein, R. & Rutter, W. J. Comparative Studies of Liver and Muscle Aldolase II. Immunochemical and Chromatographic Differentiation. *J. Biol. Chem.* **238**, 3280–3285 (1963).

113. St-Jean, M., Lafrance-Vanasse, J., Liotard, B. & Sygusch, J. High Resolution Reaction Intermediates of Rabbit Muscle Fructose-1,6-bisphosphate Aldolase. Substrate Cleavage and Induced Fit. *J. Biol. Chem.* **280**, 27262–27270 (2005).

114. Krajewska, B. Ureases I. Functional, catalytic and kinetic properties: A review. *J. Mol. Catal. B Enzym.* **59**, 9–21 (2009).

115. Weber, M., Jones, M. J. & Ulrich, J. Optimisation of isolation and purification of the jack bean enzyme urease by extraction and subsequent crystallization. *Food Bioprod. Process.* **86**, 43–52 (2008).

116. Sachs, G., Weeks, D. L., Melchers, K. & Scott, D. R. The Gastric Biology of Helicobacter pylori. *Annu. Rev. Physiol.* **65**, 349–369 (2003).

117. Balasubramanian, A. & Ponnuraj, K. Crystal Structure of the First Plant Urease from Jack Bean: 83 Years of Journey from Its First Crystal to Molecular Structure. *J. Mol. Biol.* **400**, 274–283 (2010).

118. Takishima, K., Suga, T. & Mamiya, G. The structure of jack bean urease. The complete amino acid sequence, limited proteolysis and reactive cysteine residues. *Eur. J. Biochem.* **175**, 151–157 (1988).

119. Tian, B. *et al.* Production and Characterization of a Camelid Single Domain Antibody–Urease Enzyme Conjugate for the Treatment of Cancer. *Bioconjugate Chem.* **26**, 1144–1155 (2015).

120. Walker, D., Käsdorf, B. T., Jeong, H.-H., Lieleg, O. & Fischer, P. Enzymatically active biomimetic micropropellers for the penetration of mucin gels. *Sci. Adv.* **1**, e1500501 (2015).

121. Okada, J. *et al.* Enzyme-mediated protein refolding. *Chem. Commun.* **1**, 7197 (2009).

122. Blakeley, R. L., Webb, E. C. & Zerner, B. Jack bean urease (EC 3.5.1.5). A new purification and reliable rate assay. *Biochemistry* **8**, 1984–1990 (1969).

123. Verdouw, H., Van Echteld, C. J. A. & Dekkers, E. M. J. Ammonia determination based on indophenol formation with sodium salicylate. *Water Res.* **12**, 399–402 (1978).

124. Kaltwasser, H. & Schlegel, H. G. NADH-dependent coupled enzyme assay for urease and other ammonia-producing systems. *Anal. Biochem.* **16**, 132–138 (1966).

125. Sielaff, H. *et al.* Domain compliance and elastic power transmission in rotary FOF1-ATPase. *Proc. Natl. Acad. Sci. U.S.A.* **105**, 17760–17765 (2008).

126. Yanagisawa, S. & Frasch, W. D. Protonation-dependent stepped rotation of the F-type ATP synthase c-ring observed by single-molecule measurements. *J. Biol. Chem.* **292**, 17093–17100 (2017).

127. Sobti, M. *et al.* Cryo-EM structures of the autoinhibited E. coli ATP synthase in three rotational states. *eLife* **5**, e21598 (2016).

128. Berg, J. M., Tymoczko, J. L., Stryer, L. & Clarke, N. D. *Biochemistry*. (W. H. Freeman, 2003).

129. Muench, S. P., Trinick, J. & Harrison, M. A. Structural divergence of the rotary ATPases. *Quart. Rev. Biophys.* **44**, 311–356 (2011).

130. Shah, N. B., Hutcheon, M. L., Haarer, B. K. & Duncan, T. M. F1-ATPase of Escherichia coli: the ε-inhibited state forms after ATP hydrolysis, is distinct from the ADP-inhibited state, and responds dynamically to catalytic site ligands. *J. Biol. Chem.* **288**, 9383–9395 (2013).

131. Arnold, F. H. Innovation by Evolution: Bringing New Chemistry to Life (Nobel Lecture). *Angew. Chem. Int. Ed.* **58**, 14420–14426 (2019).

132. Busch, F. *et al.* Ancestral Tryptophan Synthase Reveals Functional Sophistication of Primordial Enzyme Complexes. *Cell Chem. Biol.* **23**, 709–715 (2016).

133. Romero, P. A. & Arnold, F. H. Exploring protein fitness landscapes by directed evolution. *Nat. Rev. Mol. Cell Biol.* **10**, 866–876 (2009).

134. Hauer, B. Embracing Nature's Catalysts: A Viewpoint on the Future of Biocatalysis. *ACS Catal.* **10**, 8418–8427 (2020).

135. Faber, K. *Biotransformations in Organic Chemistry: A Textbook*. (Springer-Verlag, 2011).

136. Bar-Even, A. *et al.* The Moderately Efficient Enzyme: Evolutionary and Physicochemical Trends Shaping Enzyme Parameters. *Biochemistry* **50**, 4402–4410 (2011).

137. Wolfenden, R. & Snider, M. J. The Depth of Chemical Time and the Power of Enzymes as Catalysts. *Acc. Chem. Res.* **34**, 938–945 (2001).

138. Garcia-Galan, C., Berenguer-Murcia, Á., Fernandez-Lafuente, R. & Rodrigues, R. C. Potential of Different Enzyme Immobilization Strategies to Improve Enzyme Performance. *Adv. Synth. Catal.* **353**, 2885–2904 (2011).

139. Rosman, C. *et al.* Multiplexed Plasmon Sensor for Rapid Label-Free Analyte Detection. *Nano Lett.* **13**, 3243–3247 (2013).

140. Hogan, N. J. *et al.* Nanoparticles Heat through Light Localization. *Nano Lett.* **14**, 4640–4645 (2014).

141. Mark, A. G., Gibbs, J. G., Lee, T.-C. & Fischer, P. Hybrid nanocolloids with programmed three-dimensional shape and material composition. *Nat. Mater.* **12**, 802–807 (2013).

142. Linko, V. & Dietz, H. The enabled state of DNA nanotechnology. *Curr. Opin. Biotechnol.* **24**, 555–561 (2013).

143. Yang, Y. R., Liu, Y. & Yan, H. DNA Nanostructures as Programmable Biomolecular Scaffolds. *Bioconjugate Chem.* **26**, 1381–1395 (2015).

144. Cardinale, D., Carette, N. & Michon, T. Virus scaffolds as enzyme nano-carriers. *Trends Biotechnol.* **30**, 369–376 (2012).

145. Saccà, B. & Niemeyer, C. M. DNA Origami: The Art of Folding DNA. *Angew. Chem. Int. Ed.* **51**, 58–66 (2012).

146. Fu, J., Liu, M., Liu, Y., Woodbury, N. W. & Yan, H. Interenzyme Substrate Diffusion for an Enzyme Cascade Organized on Spatially Addressable DNA Nanostructures. *J. Am. Chem. Soc.* **134**, 5516–5519 (2012).

147. Sun, L. *et al.* Real-Time Imaging of Single-Molecule Enzyme Cascade Using a DNA Origami Raft. *J. Am. Chem. Soc.* **139**, 17525–17532 (2017).

148. Ijäs, H., Hakaste, I., Shen, B., Kostiainen, M. A. & Linko, V. Reconfigurable DNA Origami Nanocapsule for pH-Controlled Encapsulation and Display of Cargo. *ACS Nano* **13**, 5959–5967 (2019).

149. Shaw, A., Benson, E. & Högberg, B. Purification of Functionalized DNA Origami Nanostructures. *ACS Nano* **9**, 4968–4975 (2015).

150. Koch, C. *et al.* Novel roles for well-known players: from tobacco mosaic virus pests to enzymatically active assemblies. *Beilstein J. Nanotechnol.* **7**, 613–629 (2016).

151. McCafferty, J., Jackson, R. H. & Chiswell, D. J. Phage-enzymes: expression and affinity chromatography of functional alkaline phosphatase on the surface of bacteriophage. *Protein Eng. Des. Sel.* **4**, 955–961 (1991).

152. Li, K. *et al.* Chemical Modification of M13 Bacteriophage and Its Application in Cancer Cell Imaging. *Bioconjugate Chem.* **21**, 1369–1377 (2010).

153. Hwang, I. Virus Outbreaks in Chemical and Biological Sensors. *Sensors* **14**, 13592–13612 (2014).

154. Wu, F. & Minteer, S. Krebs Cycle Metabolon: Structural Evidence of Substrate Channeling Revealed by Cross-Linking and Mass Spectrometry. *Angew. Chem. Int. Ed.* **54**, 1851–1854 (2015).

155. Idan, O. & Hess, H. Diffusive transport phenomena in artificial enzyme cascades on scaffolds. *Nat. Nanotechnol.* **7**, 769–770 (2012).

156. Idan, O. & Hess, H. Origins of Activity Enhancement in Enzyme Cascades on Scaffolds. *ACS Nano* **7**, 8658–8665 (2013).

157. Zhang, Y., Tsitkov, S. & Hess, H. Proximity does not contribute to activity enhancement in the glucose oxidase–horseradish peroxidase cascade. *Nat. Commun.* **7**, 13982 (2016).

158. Sengupta, S. *et al.* Self-powered enzyme micropumps. *Nat. Chem.* **6**, 415–422 (2014).

159. Ortiz-Rivera, I., Shum, H., Agrawal, A., Sen, A. & Balazs, A. C. Convective flow reversal in self-powered enzyme micropumps. *Proc. Natl. Acad. Sci. U.S.A.* **113**, 2585–2590 (2016).

160. Valdez, L., Shum, H., Ortiz-Rivera, I., Balazs, A. C. & Sen, A. Solutal and thermal buoyancy effects in self-powered phosphatase micropumps. *Soft Matter* **13**, 2800–2807 (2017).

161. Katuri, J., Ma, X., Stanton, M. M. & Sánchez, S. Designing Micro- and Nanoswimmers for Specific Applications. *Acc. Chem. Res.* **50**, 2–11 (2017).

162. Kanti Dey, K. & Sen, A. Chemically Propelled Molecules and Machines. *J. Am. Chem. Soc.* **139**, 7666–7676 (2017).

163. Wang, X. *et al.* Fuel-Free Nanocap-Like Motors Actuated Under Visible Light. *Adv. Funct. Mater.* **28**, 1705862 (2018).

164. Krichevsky, O. & Bonnet, G. Fluorescence correlation spectroscopy: the technique and its applications. *Rep. Prog. Phys.* **65**, 251 (2002).

165. Strömqvist, J. *et al.* Quenching of Triplet State Fluorophores for Studying Diffusion-Mediated Reactions in Lipid Membranes. *Biophys. J.* **99**, 3821–3830 (2010).

166. Dertinger, T. *et al.* Two-Focus Fluorescence Correlation Spectroscopy: A New Tool for Accurate and Absolute Diffusion Measurements. *ChemPhysChem* **8**, 433–443 (2007).

167. Böttcher, B., Bertsche, I., Reuter, R. & Gräber, P. Direct visualisation of conformational changes in EF0F1 by electron microscopy. *J. Mol. Biol.* **296**, 449–457 (2000).

168. Wilkens, S. & Capaldi, R. A. Solution Structure of the ε Subunit of the F1-ATPase from Escherichia coli and Interactions of This Subunit with β Subunits in the Complex. *J. Biol. Chem.* **273**, 26645–26651 (1998).

169. Häsler, K., Pänke, O. & Junge, W. On the Stator of Rotary ATP Synthase: The Binding Strength of Subunit δ to (αβ)3 As Determined by Fluorescence Correlation Spectroscopy. *Biochemistry* **38**, 13759–13765 (1999).

170. Schmidt, K. *et al.* Design of an allosterically modulated doxycycline and doxorubicin drug-binding protein. *Proc. Natl. Acad. Sci. U.S.A.* **115**, 5744–5749 (2018).

171. Singh, D., Sielaff, H., Börsch, M. & Grüber, G. Conformational dynamics of the rotary subunit F in the A3B3DF complex of Methanosarcina mazei Gö1 A-ATP synthase monitored by single-molecule FRET. *FEBS Lett.* **591**, 854–862.

172. Dong, G. & Gregory Zeikus, J. Purification and characterization of alkaline phosphatase from Thermotoga neapolitana. *Enzyme Microb. Technol.* **21**, 335–340 (1997).

173. Bannister, A. & Foster, R. L. Buffer-Induced Activation of Calf Intestinal Alkaline Phosphate. *Eur. J. Biochem.* **113**, 199–203 (1980).

174. Paliwal, S. *et al.* Fluorescence-based sensing of p-nitrophenol and p-nitrophenyl substituent organophosphates. *Anal. Chim. Acta* **596**, 9–15 (2007).

175. Wang, W., Duan, W., Ahmed, S., Mallouk, T. E. & Sen, A. Small power: Autonomous nano- and micromotors propelled by self-generated gradients. *Nano Today* **8**, 531–554 (2013).

176. Ebbens, S. J. & Howse, J. R. In pursuit of propulsion at the nanoscale. *Soft Matter* **6**, 726 (2010).

177. Jurado-Sánchez, B. & Wang, J. Micromotors for environmental applications: a review. *Environ. Sci. Nano* **5**, 1530–1544 (2018).

178. Singh, V. V. & Wang, J. Nano/micromotors for security/defense applications. A review. *Nanoscale* **7**, 19377–19389 (2015).

179. Abdelmohsen, L. K. E. A. *et al.* Dynamic Loading and Unloading of Proteins in Polymeric Stomatocytes: Formation of an Enzyme-Loaded Supramolecular Nanomotor. *ACS Nano* **10**, 2652–2660 (2016).

180. Popescu, M. N., Uspal, W. E., Bechinger, C. & Fischer, P. Chemotaxis of Active Janus Nanoparticles. *Nano Lett.* **18**, 5345–5349 (2018).

181. Yu, H., Jo, K., Kounovsky, K. L., Pablo, J. J. de & Schwartz, D. C. Molecular Propulsion: Chemical Sensing and Chemotaxis of DNA Driven by RNA Polymerase. *J. Am. Chem. Soc.* **131**, 5722–5723 (2009).

182. Morris, K. F. & Johnson, C. S. Diffusion-ordered two-dimensional nuclear magnetic resonance spectroscopy. *J. Am. Chem. Soc.* **114**, 3139–3141 (1992).

183. Kärger, J. & Caro, J. Interpretation and correlation of zeolitic diffusivities obtained from nuclear magnetic resonance and sorption experiments. *J. Chem. Soc. Faraday Trans. 1* **73**, 1363 (1977).

184. Price, W. S. Pulsed-field gradient nuclear magnetic resonance as a tool for studying translational diffusion: Part 1. Basic theory. *Concepts Magn. Reson.* **9**, 299–336 (1997).

185. Price, W. S. Pulsed-field gradient nuclear magnetic resonance as a tool for studying translational diffusion: Part II. Experimental aspects. *Concepts Magn. Reson.* **10**, 197–237 (1998).

186. Jones, J. A., Wilkins, D. K., Smith, L. J. & Dobson, C. M. Characterisation of protein unfolding by NMR diffusion measurements. *J. Biomol. NMR* **10**, 199–203 (1997).

187. Schwarz, H. B. *et al.* In Situ 13C Fourier Transform Pulsed Field Gradient NMR Study of Intracrystalline Diffusion during Isopropanol Conversion in X-Type Zeolites. *J. Catal.* **167**, 248–255 (1997).

188. Price, W. S. & Kuchel, P. W. Effect of nonrectangular field gradient pulses in the stejskal and tanner (diffusion) pulse sequence. *J. Magn. Reson.* **94**, 133–139 (1991).

189. Sørland, G. H., Anthonsen, H. W., Zick, K., Sjöblom, J. & Simon, S. A spoiler recovery method for rapid diffusion measurements. *diffusion-fundamentals.org* **15**, 1–9 (2011).

190. Mohajerani, F., Zhao, X., Somasundar, A., Velegol, D. & Sen, A. A Theory of Enzyme Chemotaxis: From Experiments to Modeling. *Biochemistry* **57**, 6256–6263 (2018).

191. Price, W. S., Elwinger, F., Vigouroux, C. & Stilbs, P. PGSE-WATERGATE, a new tool for NMR diffusion-based studies of ligand-macromolecule binding. *Magn. Reson. Chem.* **40**, 391–395 (2002).

192. Solovev, A. A., Sanchez, S., Mei, Y. & Schmidt, O. G. Tunable catalytic tubular micro-pumps operating at low concentrations of hydrogen peroxide. *Phys. Chem. Chem. Phys.* **13**, 10131 (2011).

193. Ma, X., Hortelão, A. C., Patiño, T. & Sánchez, S. Enzyme Catalysis To Power Micro/Nanomachines. *ACS Nano* **10**, 9111–9122 (2016).

194. Rucinskaite, G., Thompson, S. A., Paterson, S. & de la Rica, R. Enzyme-coated Janus nanoparticles that selectively bind cell receptors as a function of the concentration of glucose. *Nanoscale* **9**, 5404–5407 (2017).

195. Ortiz-Rivera, I., Courtney, T. M. & Sen, A. Enzyme Micropump-Based Inhibitor Assays. *Adv. Funct. Mater.* **26**, 2135–2142 (2016).

196. Mohamad, N. R., Marzuki, N. H. C., Buang, N. A., Huyop, F. & Wahab, R. A. An overview of technologies for immobilization of enzymes and surface analysis techniques for immobilized enzymes. *Biotechnol. Biotechnol. Equip.* **29**, 205–220 (2015).

197. Kreuzer, L. P., Männel, M. J., Schubert, J., Höller, R. P. M. & Chanana, M. Enzymatic Catalysis at Nanoscale: Enzyme-Coated Nanoparticles as Colloidal Biocatalysts for Polymerization Reactions. *ACS Omega* **2**, 7305–7312 (2017).

198. Ge, J., Lei, J. & Zare, R. N. Protein–inorganic hybrid nanoflowers. *Nat. Nanotechnol.* **7**, 428–432 (2012).

199. Lei, Z. *et al.* Recent advances in biomolecule immobilization based on self-assembly: organic–inorganic hybrid nanoflowers and metal–organic frameworks as novel substrates. *J. Mater. Chem. B* **6**, 1581–1594 (2018).

200. Lin, Z. *et al.* Facile Synthesis of Enzyme-Inorganic Hybrid Nanoflowers and Its Application as a Colorimetric Platform for Visual Detection of Hydrogen Peroxide and Phenol. *ACS Appl. Mater. Interfaces* **6**, 10775–10782 (2014).

201. Wang, L.-B. *et al.* A New Nanobiocatalytic System Based on Allosteric Effect with Dramatically Enhanced Enzymatic Performance. *J. Am. Chem. Soc.* **135**, 1272–1275 (2013).

202. Koch, C. *et al.* Modified TMV Particles as Beneficial Scaffolds to Present Sensor Enzymes. *Front. Plant Sci.* **6**, (2015).

203. Hemminga, M. A. *et al.* Viruses: incredible nanomachines. New advances with filamentous phages. *Eur. Biophys. J.* **39**, 541–550 (2010).

204. Marvin, D. Filamentous phage structure, infection and assembly. *Curr. Opin. Struct. Biol.* **8**, 150–158 (1998).

205. Løset, G. Å. & Sandlie, I. Next generation phage display by use of pVII and pIX as display scaffolds. *Methods* **58**, 40–46 (2012).

206. Rothenstein, D. *et al.* Mineralization of gold nanoparticles using tailored M13 phages. *Bioinspired Biomim. Nanobiomaterials* **2**, 173–185 (2013).

207. Kilper, S. *et al.* Macroscopic Properties of Biomimetic Ceramics Are Governed by the Molecular Recognition at the Bioorganic-Inorganic Interface. *Adv. Funct. Mater.* **28**, 1705842 (2018).

208. Kilper, S. *et al.* Genetically Induced In Situ-Poling for Piezo-Active Biohybrid Nanowires. *Adv. Mater.* **31**, 1805597 (2018).

209. Adhikari, M. *et al.* Functionalized viral nanoparticles as ultrasensitive reporters in lateral-flow assays. *Analyst* **138**, 5584 (2013).

210. Adhikari, M. *et al.* Aptamer-Phage Reporters for Ultrasensitive Lateral Flow Assays. *Anal. Chem.* **87**, 11660–11665 (2015).

211. Lee, Y. J. *et al.* Fabricating Genetically Engineered High-Power Lithium Ion Batteries Using Multiple Virus Genes. *Science* **324**, 1051–1055 (2009).

212. Nam, K. T. *et al.* Stamped microbattery electrodes based on self-assembled M13 viruses. *Proc. Natl. Acad. Sci. U.S.A.* **105**, 17227–17231 (2008).

213. Yoo, S. Y., Chung, W.-J. & Lee, D.-Y. Chemical modulation of M13 bacteriophage and its functional opportunities for nanomedicine. *IJN* **9**, 5825 (2014).

214. Patiño, T. *et al.* Influence of Enzyme Quantity and Distribution on the Self-Propulsion of Non-Janus Urease-Powered Micromotors. *J. Am. Chem. Soc.* **140**, 7896–7903 (2018).

215. Müller, M. Chemoenzymatic Synthesis of Building Blocks for Statin Side Chains. *Angew. Chem. Int. Ed.* **44**, 362–365 (2004).

216. Smith, G. P. Absorption Spectrum and Quantitation of Filamentous Phage. http://www.biosci.missouri.edu/smithgp/ PhageDisplayWebsite/PhageDisplayWebsiteindex.html.

217. Zhang, Y. & Hess, H. Enhanced Diffusion of Catalytically Active Enzymes. *ACS Cent. Sci.* **5**, 939–948 (2019).

218. Feng, M. & Gilson, M. K. Enhanced Diffusion and Chemotaxis of Enzymes. *Annu. Rev. Biophys.* **49**, 87–105 (2020).

219. Chen, Z. *et al.* Single-Molecule Diffusometry Reveals No Catalysis-Induced Diffusion Enhancement of Alkaline Phosphatase as Proposed by FCS Experiments. *Proc. Natl. Acad. Sci. U.S.A.* **117**, 21328–21335 (2020).

220. Oikonomou, M., Asencio-Hernández, J., Velders, A. H. & Delsuc, M.-A. Accurate DOSY Measure for Out-of-Equilibrium Systems Using Permutated DOSY (p-DOSY). *J. Magn. Reson.* **258**, 12–16 (2015).

221. Urbańczyk, M., Bernin, D., Czuroń, A. & Kazimierczuk, K. Monitoring Polydispersity by NMR Diffusometry with Tailored Norm Regularisation and Moving-Frame Processing. *Analyst* **141**, 1745–1752 (2016).

222. Ghosh, S. *et al.* Motility of Enzyme-Powered Vesicles. *Nano Lett.* **19**, 6019–6026 (2019).

223. Hoggett, J. G. & Kellett, G. L. Yeast Hexokinase: Substrate-Induced Association- Dissociation Reactions in the Binding of Glucose to Hexokinase P-II. *Eur. J. Biochem.* **66**, 65–77 (1976).

224. Furman, T. C. & Neet, K. E. Association equilibria and reacting enzyme gel filtration of yeast hexokinase. *J. Biol. Chem.* **258**, 4930–4936 (1983).

225. Kadiri *et al.* Genetically Modified M13 Bacteriophage Nanonets for Enzyme Catalysis and Recovery. *Catalysts* **9**, 723 (2019).

226. Feynman, R. P. There's Plenty of Room at the Bottom. *Eng. Sci.* **23**, 22–36 (1960).

227. Covington, A. K., Paabo, Maya., Robinson, R. Anthony. & Bates, R. G. Use of the glass electrode in deuterium oxide and the relation between the standardized pD (paD) scale and the operational pH in heavy water. *Anal. Chem.* **40**, 700–706 (1968).

228. Gill, S. C. & von Hippel, P. H. Calculation of protein extinction coefficients from amino acid sequence data. *Anal. Biochem.* **182**, 319–326 (1989).

229. Saavedra, E. *et al.* Kinetic modeling can describe in vivo glycolysis in Entamoeba histolytica: Modeling Entamoeba glycolysis. *FEBS J.* **274**, 4922–4940 (2007).

230. Goldberg, R. N. & Tewari, Y. B. Thermodynamics of Enzyme-Catalyzed Reactions: Part 4. Lyases. *J. Phys. Chem. Ref. Data* **24**, 1669–1698 (1995).

231. Callens, M., Kuntz, D. A. & Opperdoes, F. R. Kinetic properties of fructose bisphosphate aldolase from Trypanosoma brucei compared to aldolase from rabbit muscle and Staphylococcus aureus. *Mol. Biochem. Parasitol.* **47**, 1–9 (1991).

232. Rutter, W. J., Woodfin, B. M., Blostein, R. E. & Baptist, J. N. Enymic Homology. Structural and Catalytic Differentiation of Fructose Diphosphate Aldolase. *Acta Chem. Scand.* **17 supl.**, 226–232 (1963).

233. Midelfort, C. F., Gupta, R. K. & Rose, I. A. Fructose 1,6-bisphosphate: isomeric composition, kinetics, and substrate specificity for the aldolases. *Biochemistry* **15**, 2178–2185 (1976).

234. Qi, D. & Scholthof, K.-B. G. A one-step PCR-based method for rapid and efficient site-directed fragment deletion, insertion, and substitution mutagenesis. *J. Virol. Methods* **149**, 85–90 (2008).

235. Morag, O., Sgourakis, N. G., Baker, D. & Goldbourt, A. The NMR–Rosetta capsid model of M13 bacteriophage reveals a quadrupled hydrophobic packing epitope. *Proc. Natl. Acad. Sci. U.S.A.* **112**, 971–976 (2015).

Acknowledgements

I would like to use this opportunity to thank the many people who helped me on this long path to the completion of this thesis. Since I met so many nice people throughout the last years, I would like to apologize in case I forgot to mention somebody.

First, I would like to thank my supervisor *Peer Fischer*. It can hardly be overstated how good *Peer* is as a doctoral adviser and scientist. He has been overwhelmingly supportive of me and my scientific endeavors and never let me down. During the last four years, *Peer's* door was always open when I needed help or advice. He opened up countless opportunities for me to grow as a scientist and did not only give me access to all the equipment I needed, but also sent me to the right places, where I met many nice and helpful people. *Peer* was and still is a highly inspirational mentor to me. Thank you, *Peer*.

Michael Börsch and *Günter Majer* both introduced me to the fascinating world of diffusion measurements – whether in absolute darkness to see every single photon with one's own eyes or with precious deuterated solvents next to a superconducting magnet. Thank you very much, *Michael* and *Günter*, for teaching me so much and for always being patient with me.

I would also like to thank our international collaborators, especially *Henry Hess* and his laboratory in New York, where I had a fantastic time, learned a lot about the American research culture and met very nice people. Thank you, *Henry*, for being an amazing host. Moreover, I would also like to thank our very recently joined collaboration with the laboratories of *Jon Beves* and *Bill Price* in Australia. I look forward to our ongoing scientific endeavors.

Of course, it is not possible to do research alone. I would therefore like to thank all the people with whom I worked together. Special thanks go to *Mariana Alarcón-Correa* for introducing me to the lab and the countless times she helped me out, *Johannes Sachs* for fascinating experiments and good physics talk, *Dirk Rothenstein* for our joint work and for providing us with phages, *Jonas Troll* for his help with experiments throughout his bachelor thesis and beyond, and *Andrew Mark* and *Jeong Hyeon-Ho* for inspirational conversations.

Additionally, I would like to thank in no particular order *Vincent, Kai, Rahul, Eunjin, Hannah, Udit, Florian, Jutta,* and all the past and present members of the Micro, Nano, & Molecular Systems Lab for the fertile and pleasant environment and support during good and bad times.

Finally, I would like to thank my mother *Evi*, my father *Horst* and my brother *Julian*, who have supported me during my whole life and have been very patient with me. I thank you for everything that you have done for me and for all that love.

Declaration of Authorship

I hereby certify that the dissertation entitled

Advanced Diffusion Studies of Active Enzymes and Nanosystems

is entirely my own work except where otherwise indicated. Passages and ideas from other sources have been clearly indicated.

Erklärung über die Eigenständigkeit der Dissertation

Ich versichere, dass ich die vorliegende Arbeit mit dem Titel

Advanced Diffusion Studies of Active Enzymes and Nanosystems

selbständig verfasst und keine anderen als die angegebenen Quellen und Hilfsmittel benutzt habe. Aus fremden Quellen entnommene Passagen und Gedanken sind als solche kenntlich gemacht.

Name/Name: Jan-Philipp Günther

Date/Datum: 27 January 2021

www.ingramcontent.com/pod-product-compliance
Ingram Content Group UK Ltd.
Pitfield, Milton Keynes, MK11 3LW, UK
UKHW021652190726
13853UKWH00001B/219